T0332347

Green Photonics
and Smart Photonics

RIVER PUBLISHERS SERIES IN OPTICS AND PHOTONICS

Volume 1

Series Editors

Manijeh Razeghi
Northwestern University
USA

Kevin Williams
Eindhoven University of Technology
The Netherlands

Advisor
Haiyin Sun
ChemImage Corp
USA

The "River Publishers Series in Optics and Photonics" is a series of comprehensive academic and professional books which focus on the theory and applications of optics, photonics and laser technology.

Books published in the series include research monographs, edited volumes, handbooks and textbooks. The books provide professionals, researchers, educators, and advanced students in the field with an invaluable insight into the latest research and developments.

Topics covered in the series include, but are by no means restricted to the following:

- Integrated optics and optoelectronics
- Applied laser technology
- Lasers optics
- Optical Sensors
- Optical spectroscopy
- Optoelectronics
- Biophotonics photonics
- Nano-photonics
- Microwave photonics
- Photonics materials

For a list of other books in this series, visit www.riverpublishers.com

Green Photonics and Smart Photonics

Editors

Shien-Kuei Liaw

National Taiwan University of Science and Technology
Taiwan

Gong-Ru Lin

National Taiwan University
Taiwan

River Publishers

Published, sold and distributed by:
River Publishers
Alsbjergvej 10
9260 Gistrup
Denmark

River Publishers
Lange Geer 44
2611 PW Delft
The Netherlands

Tel.: +45369953197
www.riverpublishers.com

ISBN: 978-87-93379-27-5 (Hardback)
      978-87-93379-26-8 (Ebook)

©2016 River Publishers

Contents

Preface

The chapters in this edited book are written by some authors who have presented high-quality papers at the 2015 International Symposium of Next-Generation Electronics (ISNE 2015) held in Taipei, Taiwan. The ISNE 2015 was intended to providea common forum for researchers, scientists, engineers, and practitioners throughout the world to present their latest research findings, ideas, developments, and applications in the general areas of electronic/photonics devices, systems, and technologies. The scope of the conference includes the following topics:

A. Green Electronics
B. Microelectronic Circuits and Systems
C. Integrated Circuits and Packaging Technologies
D. Computer and Communication Engineering
E. Electron Devices
F. Optoelectronic and Semiconductor Technologies

The technical program consists of 5 plenary talks, 23 invited talks, and more than 250 contributed oral and poster presentations. After a rigorous review process, the ISNE 2015 technical program committee has selected 10 outstanding presentations and invited the authors to prepare extended chapters for inclusion in this edited book. Of the 10 chapters, 5 focus on the subject of *green photonics* and the others cover *smart photonics*. As the guest editors, we would like to express our sincere gratitude to all the members of the ISNE 2015 technical program committees for reviewing the papers and selecting the manuscripts for the edited book. We also thank the authors for their valuable and excellent contributions to the book.

Guest Editors:
Shien-Kuei Liaw
National Taiwan University of Science and Technology, Taiwan
Gong-Ru Lin
National Taiwan University, Taiwan

List of Contributors

Liang-Chiun Chao, *Department of Electronic and Computer Engineering, National Taiwan University of Science and Technology, Taipei, Taiwan 106*

Jingo Chen, *National Center for Research on Earthquake Engineering, Taiwan*

Ku-Hung Chen, *Institute of Photonics Technologies and Department of Electrical Engineering, National Ting-Hua University, Hsinchu, Republic of China*

Wei-Ting Chen, *Institute of Photonics Technologies and Department of Electrical Engineering, National Ting-Hua University, Hsinchu, Republic of China*

Yen-Yu Chen, *Department of Photonics, National Cheng Kung University, Tainan 70101, Taiwan*

Chi-Hsien Cheng, *Graduate Institute of Photonics and Optoelectronics, and Department of Electrical Engineering, National Taiwan University, Taipei, 10617, Taiwan, Republic of China*

Yu-Chieh Chi, *Graduate Institute of Photonics and Optoelectronics, and Department of Electrical Engineering, National Taiwan University, Taipei, 10617, Taiwan, Republic of China*

Ching-Hsiang Chiu, *Institute of Photonics Technologies and Department of Electrical Engineering, National Ting-Hua University, Hsinchu, Republic of China*

Koonchun Lai, *University Tunku Abdul Rahman, Kampar, Malaysia*

Ming-Chang M. Lee, *Institute of Photonics Technologies and Department of Electrical Engineering, National Ting-Hua University, Hsinchu, Republic of China*

Ming-Yi Lee, *Parallel and Scientific Computing Laboratory, Institute of Communications Engineering, Department of Electrical and Computer Engineering, National Chiao Tung University, Hsinchu 300, Taiwan*

Yiming Li, *Parallel and Scientific Computing Laboratory, Institute of Communications Engineering, Department of Electrical and Computer Engineering, National Chiao Tung University, Hsinchu 300, Taiwan*

Shien-Kuei Liaw, *National Taiwan University of Science and Technology, Taiwan*

Gong-Ru Lin, *Graduate Institute of Photonics and Optoelectronics, and Department of Electrical Engineering, National Taiwan University, Taipei 10617, Taiwan, Republic of China*

Yung-Hsiang Lin, *Graduate Institute of Photonics and Optoelectronics, and Department of Electrical Engineering, National Taiwan University, Taipei 10617, Taiwan, Republic of China*

Wen-Fung Liu, *Department of Electrical Engineering, Feng Chia University, 100 Wenhwa Road, Seatwen, Taichung, Taiwan 407, Republic of China*

Cheemeng Loo, *University Tunku Abdul Rahman, Kampar, Malaysia*

Neil Na, *Institute of Photonics Technologies and Department of Electrical Engineering, National Ting-Hua University, Hsinchu, Republic of China*

Kokseng Ong, *University Tunku Abdul Rahman, Kampar, Malaysia*

Seiji Samukawa, *Institute of Fluid Science and WPI-AIMR, Tohoku University, Sendai 980-8577, Japan*

Hao-Jan Sheng, *Department of Electrical Engineering, Feng Chia University, 100 Wenhwa Road, Seatwen, Taichung, Taiwan 407, Republic of China*

Choonfoong Tan, *University Tunku Abdul Rahman, Kampar, Malaysia*

Yi-Chia Tsai, *Parallel and Scientific Computing Laboratory, Institute of Communications Engineering, Department of Electrical and Computer Engineering, National Chiao Tung University, Hsinchu 300, Taiwan*

Chih-Kuo Tseng, *Institute of Photonics Technologies and Department of Electrical Engineering, National Ting-Hua University, Hsinchu, Republic of China*

Chung-Lun Wu, *Graduate Institute of Photonics and Optoelectronics, and Department of Electrical Engineering, National Taiwan University, Taipei 10617, Taiwan, Republic of China*

Wenchang Yeh, *Interdisciplinary Graduate School of Science and Engineering, Shimane University, Matsue, Shimane 690-8504, Japan*

Shih-Che Yen, *Institute of Photonics Technologies and Department of Electrical Engineering, National Ting-Hua University, Hsinchu, Republic of China*

Yi-Lin Yu, *National Taiwan University of Science and Technology, Taiwan*

List of Figures

List of Tables

List of Abbreviations

Al	Aluminum
AlN	Aluminum nitride
ASE	Amplified spontaneous emission
a-SiC	Amorphous SiC
AWG	Arbitrary waveform generator
BN	Baron nitride
BOE	Buffered oxide etchant
BOX	Buried oxide
CFD	Computational fluid dynamics
CB	Conduction band
CVD	Chemical vapor deposition
Cat-CVD	Catalytic chemical vapor deposition
C	Carbon
DoS	Density of states
DFA	Doped fiber amplifier
DCM	Dispersion compensation module
DP	Differential pressure
DC	Dark current
EQE	External quantum efficiency
EDFA	Erbium-doped fiber amplifier
EYDF	Er/Yb co-doped fiber
EDS	Energy dispersive spectrometer
EL	Electroluminescence
FEM	Finite element method
FWHM	Full-width at half maximum
FGA	Forming gas annealing
FF	Filling factor
FBG	Fiber Bragg grating
FDTD	Finite-difference-time-domain
FCA	Free-carrier absorption effects
FCD	Free-carrier plasma dispersion

FIB	Focused ion beam
GaN	Gallium nitride
GaAs	Gallium arsenide
GVD	Group velocity dispersion
GeSn	Germanium-tin
HR	High resolution
HWCVD	Hot-wire chemical vapor deposition
HRTEM	High-resolution transmission electron microscopy
IBSC	Intermediate band solar cell
ITO	Indium tin oxide
IPCE	Incident photon-to-current conversion efficiency
ISO	Isolators
ILD	Inter-layer dielectric
ICP	Inductively coupled plasma
LED/LEDs	Light-emitting diodes
LD	Laser diode
LPCVD	Low-pressure chemical vapor deposition
LMIS	Liquid metal ion source
MBE	Molecular beam epitaxy
MOCVD	Metal organic chemical vapor deposition
MOVPE	Metal organic vapor phase epitaxy
MSM	Metal–semiconductor–metal
MEMS	Micro-electro mechanical system
NBE	Neutral beam etching
NHE	Normal hydrogen electrode
NRZ	Non-return-to-zero
NLSE	Non-linear Schrodinger equation
NRZ-OOK	Non-return-to-zero on-off-keying
NBE	Near band edge
1D	One-dimensional
OC	Optical circulator
PCB	Printed circuit board
PECVD	Plasma-enhanced chemical vapor deposition
Pt	Platinum
PEC	Photoelectrochemical
PBC	Polarization beam combiner
PDG	Polarization-dependent gain
PC	Photocurrent
PL	Photoluminescence

KCl	Potassium chloride
QDs	Quantum dots
RHEED	Reflective high-energy electron diffraction
RMG	Rapid melt growth
RTA	Rapid thermal annealing
RIE	Reactive-ion etching
RMS	Root mean square
SE	Sputtering epitaxy/sputter epitaxy
SCs	Solar cells
SIMS	Secondary ion mass spectroscopy
SiC	Silicon carbide
SOA	Semiconductor optical amplifier
SMF	Single-mode fiber
SAMB	Self-assembled microbonding
SAD	Selected-area diffraction
SOI	Silicon-on-insulator
SRH	Schockey–Read–Hall
Si_3N_4	Silicon nitride
Si-QD	Si quantum-dot
TIM	Thermal interface material
TEC	Thermoelectric cooler
tp	Thermal paste
3D	Three-dimensional
2D	Two-dimensional
TEM	Transmission electron microscope
TDD	Threading dislocation density
TOD	Third-order dispersion
TIA	Trans-impedance amplifiers
TPA	Two-photon absorption
UV	Ultraviolet
VB	Valence band
VLS	Vapor–liquid–solid
VS	Vapor–solid
WDM	Wavelength division multiplexer
XPS	X-ray photoelectron spectroscopy
ZnCl	Zinc chloride

Introduction

In recent years, much effort has been devoted to the study, development, and application of green photonics and smart photonics. This book presents results of recent advances made both in theory and applications to reflect cutting-edge technologies and research achievements in green photonics and smart photonics.

The aim of green photonics is to develop photonics technologies that can generate or conserve energy, cut greenhouse gas emissions, reduce pollution, create renewable energy, and yield environmentally sustainable outputs. Green photonics is a key technology for achieving a global balance in the levels of atmospheric carbon dioxide and it will become very important in the near future. Also, green photonics can be used in solid-state lighting, photovoltaic solar cells (PVSCs), and optical communications. In Chapters 1–5 of this book, LEDs and PVSCs having the characteristics of sustainable and low energy consumption will be addressed separately.

In Chapter 1, Koonchun Lai et al. introduce the "Thermal field study of multichip LED module." They discuss the thermal dissipation technique of multichip packaged LEDs for cost-effective lighting applications and the thermal transfer model to predict the temperature gradient uniformity, which could inevitably induce unbalanced stress along the multichip LED interface to cause output degradation.

In Chapter 2, Lee et al. discuss the "Modeling and simulation of Ge/Si nano-disk array for QD-based IBSCs." The finite element simulation is performed to calculate the band structure and DoS for well-ordered Ge/Si nanodisk array fabricated by self-assembly and NBE. Within the envelop-function framework, their model surmounts theoretical approximations of the multidimensional Kronig-Penney method and accurately calculates the energy dispersion relationship. The observation is helpful for the design of intermediate-band PVSCs with 3DQDs.

In Chapter 3, the "SE of Si and Ge for application to PVSCs" is investigated by Yeh. Owing to the need for massive production, a low-cost and large-area synthesis method such as sputtering is employed for fabricating

1

Si- or Ge-based PVSCs, and a multi junction Si/SiGe-Ge PVSC is performed. By co-sputtering the acceptors or donors to control conductivity, then-i-p thin-film PVSCs on Si(100) or Ge(100) wafer are successfully demonstrated.

In Chapter 4, the "Non-stoichiometric SiC-Based PVSCs" are investigated by Cheng et al. The C/Si composition ratio of the non-stoichiometric Si_xC_{1-x} film can be detuned under the PECVD synthesis, which exhibits a high optical absorption coefficient at visible wavelength region. The intrinsic Si_xC_{1-x} layer is added to enhance the conversion efficiency, and the series resistance of the non-stoichiometric Si_xC_{1-x} film is optimized. The overall performance can be further improved by combing the amorphous Si (a-Si)-based p-i-n junction PVSC to form the tandem PVSC.

In Chapter 5, Chen et al. study "Water splitting using GaN-based working electrodes for hydrogen generation with bias by PVSCs." PEC hydrogen generation by GaN at working electrode for water splitting under solar illumination, the efficiency of hydrogen generation can be improved by increasing bias at working electrode. Instead of using external bias, a PVSC can be used as the driving force, and the GaN is suitable for photo-electrolysis. Hybrid working electrodes, including n-GaN and InGaN, can also be employed to improve the efficiency of hydrogen generation.

The term "smart photonics" reflect intelligence of optical and optoelectronic components with good sensitivity, high response time, and/or compact size. There are many examples of smart photonics such as smart photonic coating, smart photonic cloud, smart photonic networking, smart structure for photonic integration, smart light source, optoelectronic integrated circuits, and optical sensing. From Chapters 6 to 10, the various aspects of smart photonics, including smart light source and amplifier, fiber sensor, optoelectronic device, and waveguide device, are explored. The subject matter and abstract of each chapter is mentioned below in sequence.

In Chapter 6, Yu et al. present several types of fiber amplifier which could be used in communication and sensing, including EDFA, RFA, hybrid fiber amplifier (EDFA+RFA), and high-power fiber amplifier. The authors also used the residual pumping power to improve the gain efficiency of C+L band signals.

In Chapter 7, Liu et al. highlight several sensing applications using FBGs. For examples, a lateral pressure sensor by combining the structure packaging design with pressure sensitivity much more higher than that of using a bare FBG, high-sensitivity temperature-independent DP sensor of sensitivity 5.27×10^{-1} Mpa^{-1}, and a random rotational angle sensor using two fiber gratings with high accuracy for detecting the rotary position with full around 360° in any direction of a rotor, etc.

In Chapter 8, Tseng et al. present several waveguide-based high-speed Si/Ge/Sn photo detectors, including Si/Ge hetero junction waveguide pin, Si/Ge butt-coupling waveguide photo detectors, and GeSn photo detectors. Also, the RMG method, in combination with the self-aligned microbonding technique, is applied to heterogeneously integrate the mono-crystalline Group IV semiconductor on Si photonic devices. This is a unique process.

In Chapter 9, Wu et al. demonstrate an ultrafast nonlinear optical Kerr switch with Si-QD doped in an a-SiC (a-SiC:Si-QD) micro-ring resonator. The optical non-linearity of a-SiC is significantly enhanced by the enlarged oscillation strength of localized excitons in the Si-QD. TPA can be significantly suppressed by setting the operation wavelength at 1550 nm, which is made possible by the small photon energy. Such a property is very important to analyze the nonlinear Kerr switch at telecommunication wavelengths without interfering with the TPA and FCA effect.

In the final Chapter, Chao discusses the Ion beam, a fundamental tool in the semiconductor industry that has been utilized in various vital applications. Broad beam ion sources are commonly used in surface cleaning, material deposition, and dopant implantation. Besides, because of the invention of high brightness liquid metal ion sources, FIB work station has also become a standard tool in nanofabrication and integrated circuit repair.

1

Thermal Field Study of Multichip LED Module

Koonchun Lai, Kokseng Ong, Cheemeng Loo and Choonfoong Tan

University Tunku Abdul Rahman, Kampar, Malaysia

Abstract

To date, LED was often manufactured as a multi-package module instead of as a single package keeping in mind the luminosity demand and cost concerns. Nevertheless, the uniformity of the temperature gradient may generate an unbalanced stress along the multi-package interface, resulting in lifespan and output degradation. It is hence obligatory to predict and model the thermal transfer performance of the LED module prior to the manufacturing process so as to ensure output quality and to prevent catastrophic failure.

Keywords: Light-emitting diode, computational fluid dynamics, thermo-electric cooler.

1.1 Introduction

For decades, LED has been widely used in a number of applications such as backlighting for liquid crystal display (Williams et al., 2007), street lamps (Luo et al., 2007), head light lamp for automobiles (Steranka et al., 2002), and advertising signage (Yamamoto et al., 2010). LED has drawn public attention for its better reliability and lifespan when compared to the traditional tungsten and florescence lighting. In order to enhance the illumination performance of LED, the requirement of driving current is crucial in order to produce more lumen. However, most of the input power is lost as heat before transforming into useful light (Cheng et al., 2012). The dissipated heat increases the

Green Photonics and Smart Photonics, 5–24.

temperature of the LED and affects its reliability and durability (Choi and Shin, 2012). The rise of temperature gradient generates undesired stresses in LED and eventually leads to light output degradation and catastrophic failure. Thus, thermal management is important for maintaining the output quality and performance reliability of the LED. This can be achieved by limiting the temperature gradient from exceeding the maximum rated chip temperature through proper designing and packaging.

To date, LED packages are mounted as single chip and multiple chip modules. The multichip module typically consists of more than one LED package so as to obtain more lumen at a cheaper cost. Nevertheless, the uniformity of the temperature gradient may change and generate imbalanced stress along the multichip interface. In line with this, thermal spreading analysis shall be carried out in order to predict and model the heat transfer performance of LED. Thermal distribution on single package and multiple-package modules has become a new study (Cheng et al., 2010).

1.1.1 Background of LED

LED has received a lot of attention recently. There is the possibility that LED will soon become a standard illumination source in the construction and agriculture sectors. By increasing the input driving current and the number of chips of the LED package, a large amount of luminous flux can be produced. However, it causes the junction temperature of the chip to rise and the excessive heat may degrade the performance of the LED package. High junction temperature induces the defect of electron-hole recombination and results in a decline in the total light intensity (Chuang et al., 1997). Consequently the lifespan is shortened and the output wavelength is amended undesirably. This problem of reliability poses a challenge to researchers in formulating optimum thermal management strategies (Meneghini et al., 2010; Perpiñà et al., 2012).

Thermal resistance plays a key role in the performance and reliability of LED package. High thermal resistance implies high junction temperature and causes a low lighting efficiency. Some references point out the relationship between the junction temperature and possible factors such as substrate thickness, size of the heat slug, and shape of the package mold (Darwish et al., 2004; Ha and Grahan, 2012; Liu et al., 2011; Ma and Zheng, 2007, Yang et al., 2007). Therefore, effective design and thermal management are crucial for improvement of LEDs.

CFDs is an approach to model and envisage problems related to gases, liquids, solids, and even those pertaining to multiphase. The CFD analysis enables greater robustness for reducing ambiguity in the design process, enhancing productivity, and consequently churning out higher-quality products. It measures and analyzes temperature flows and heat transfer for better thermal management. With the aid of CFD, a suitable model for LED can be identified under various heating and cooling conditions. Nevertheless in terms of heat dissipation, heat sinks are the most common hardware component in control of thermal management in electronics. With the use of fins which enhance the surface area of the heat sinks, responsiveness and effectiveness of thermal control of electronic components can be achieved. Applications utilizing fins of the heat sinks for cooling purposes have increased significantly during the last few decades due to an increase in heat flux densities and product miniaturization. Theoretical study of the heat sinks will be described and the efficacy of heat sinks in thermal reduction will be illustrated in the following sections.

1.1.2 Literatures

Back in 1995, Linton compared the results of a detailed CFD model of a heat sink with sets of experimental data. He developed a technique to represent the heat sink with a specific CFD model and the result indicated that the technique can be used for designing heat sinks for larger card or system models. Narasimhan et al. (2003) reduced the computational method by creating a heat sink using a block model with a fine pressure loss coefficient and thermal conductivity. From the study, parallel plate heat sinks arranged in a laminar shape will impose convection. This method determined the thermal characteristics of compact heat sink models with a high level of precision. The results proved that the data generated from using a porous block compact model are legit to predict the thermal characteristics of the heat sinks. Poppe et al. (2008) discussed thermal characterization of packaged semiconductor devices by performing thermal simulations. They, however, pointed out the difficulties in creating ideal simulation models since the actual time-constants of heat sink are small.

Farkas et al. (2005) proved that the structure function method can be used to identify the heat conductance path and determine the thermal resistances. They also suggested designing a compact LED model with different complexity levels to describe the measured thermal effects which lead to electrothermal simulation. The design model reflects the changes in thermal resistance and temperature transients under different current inputs. On the other hand, Tsai et al. (2012) simulated the temperature distribution by referring to the spherical coordinate system change in heat resistance. They attempted to identify the problems that affect the temperature distributions of LEDs. The simulation result confirmed the effectiveness and accuracy of using thermal resistance calculations in predicting temperature profiles. Similarly, Pan et al. (2014) determined the thermal resistance by finite element simulation on the basis of the thermal resistance network model and the concept of steady-state heat transfer. A small variance of 3.9% in terms of total thermal resistance, comparing both the experimental result and finite element simulation, was obtained.

For heat dissipation, various methods were suggested to enhance thermal performance. Li et al. (2009) analyzed the heat dissipation with a water miniature heat pipe on the LED packaging structure. The junction temperature of the source was below 70°C at a condition of natural convection. The result suggested that the heat pipe is an effective solution for the heat dissipation of LED. The desired junction temperature profiles could be achieved by altering the combination of some optimization levels in terms of height and thickness. Besides, Anderson et al. (2013) employed CFD simulations to simulate fluid and heat flow performance of a package through large power dissipations of 1000 W/m^2. A combination of using high thermal conductivity materials and forced convection cooling with DC fans could accomplish an ideal thermal control solution. Recently, Ashry (2015) made an attempt to prove the theoretical model by designing a heat sink with a lower temperature gradient than the excess heat on the PCB. The key process in heat sink design was to sample the parts situated in central areas within the circuit board for better direct access to the heat source. Experiments showed that conductive resistance on the surface of the heat sink decreased with an increased convective air flow velocity. This chapter discusses the approaches of numerical modeling by using CFD software to investigate the optimal thermal performance. Meanwhile, empirical evaluations on the LED package will be described to serve as a verification of results.

1.2 Structure and Properties of LED Package

Typically, a LED package comprises chip, die attach, lead frame, slug, lens, and solder paste on a PCB. The chip was attached to the slug by epoxy. The package was later attached to the heat sink with TIM. Aluminum was typically used for heat sink. The parameters used in modeled packages should be corresponded with the real packages. A simple schematic diagram of a LED configuration is shown in Figure 1.1. The heat is produced by the heat element in the chip and later dissipated through the slug and heat sink. Regarding the one-directional heat flow path from the chip to the heat sink, the effects of the lead frame and gold wire were usually not considered in the calculation.

The epoxy lens has a low thermal conductivity of 0.68 W/mK. It was sometimes ignored during the simulation process. The chip is in general made of GaN or GaAs. The former is a blue chip, whereas the latter appears as red. Thermal conductivities of the common packaging materials are summarized in Table 1.1.

A number of assumptions are to be made in order to simplify the simulation process. First, we assume that the temperature change during the experiment

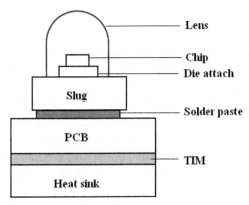

Figure 1.1 Schematic diagram of LED package.

Table 1.1 Thermal conductivity of materials

Properties	Material	Thermal Conductivity (W/mK)
Chip	GaAs	48.4
	GaN	130–140
Die attach	Silicon	0.2
	Silver	57
Slug	Copper	385–407
Solder paste	Silver (Ag)	7.5
	Sn-3.5Ag	33
	SnAgCu	51
	Au-20Sn	57
Substrate	Metal-core PCB	150–170
TIM	Grease	2–3
Heat sink	Al	150–201

is relatively small and thus the thermal conductivities of materials are independent of temperature. Besides, we may assume that the heat dissipation by convection around the die to be zero due to the negligible contacting surface areas. A uniform heat transfer coefficient, h around the heat sink can be assumed as 10–20 W/m^{2}K at 25°C ambient temperature. As most of the generated heat, about 90%, is dissipated through the heat sink, we assume that the surfaces of the substrate are adiabatic. For the boundary condition between the substrate and the TIM, a uniform temperature distribution at the bottom of TIM is assumed due to the very minor variation (<1°C). This is possible according to the low thermal conductivity of TIM.

1.3 Theoretical Studies

1.3.1 Heat Transfer

As previously mentioned in literatures, heat transfer characteristics of the LED package can be analyzed by the structure function method. It discusses the heat capacity parameters and thermal conductivity of the properties along the one-directional heat flow path. In order to derive the structure function, the thermal evaluations were first identified, followed by the heating or cooling curves. By transforming the curves, the structure function can be determined. The function relates to a time-constant system, in which the time constant, τ is a function of thermal resistance R and thermal capacitance C. According to Székely (1991), detailed response of the structure function is calculated as

$$\Delta T_j(t) = P_{\text{th}} \sum_{i=1}^{n} R_{\text{thi}} \cdot [1 - \exp(-t/\tau_i)], \qquad (1.1)$$

where P_{th} is the heat power and R_{th} is the resistance. The function approaches infinity in a real distributed system, and Equation (1.1) can be rewritten as

$$\Delta T_j(t) = P_{\text{th}} \int_0^\infty R(\tau) [1 - \exp(-t/\tau)] \, d\tau. \qquad (1.2)$$

In relation to the time constant spectrum, Equation (1.2) can be viewed as

$$\frac{d}{dz} \Delta T_j(z) = P_{\text{th}} \int_{-\infty}^\infty R(\xi) [\exp(z - \xi - \exp(z - \xi))] \, d\xi, \qquad (1.3)$$

where $z = \ln(t)$ and $\xi = \ln(\tau)$. The derivative of z is then expressed as

$$\frac{d}{dz}\Delta T_j (z) = P_{th} \cdot R (z) \otimes W (z),\qquad(1.4)$$

where $W(z) = \exp[z - \exp(z)]$. The symbol $\otimes$ denotes the convolution operation. Besides, the structure function can be represented by a distributed RC thermal system, in which the sum of thermal capacitances, and the sum of thermal resistances are derived as Equations (1.5) and (1.6), respectively:

$$C_{\Sigma} = \int_0^x c (\xi) A (\xi) \, d\xi \qquad(1.5)$$

$$R_{\Sigma} = \int_0^x \frac{d\xi}{\lambda (\xi) A (\xi)} \qquad(1.6)$$

The differential structure function can be, therefore, determined as

$$Q (R_{\Sigma}) = \frac{dC_{\Sigma}}{dR_{\Sigma}} = c(x)\lambda(x)A^2(x),\qquad(1.7)$$

where $c(x)$ is the volumetric heat capacitance, $\lambda(x)$ is the thermal conductivity, and $A(x)$ is the cross-sectional area of the heat flow.

1.3.2 Thermal Resistance

In order to determine the thermal resistance (R_{th}), the schematic of the LED package shown in Figure 1.1 can be transformed into the thermal circuit, as shown in Figure 1.2.

For the single chip LED package, the thermal resistance for the entire system can be estimated as

$$R_{tot} = R_{die} + R_{die\text{-}attach} + R_{slug} + R_{solder\text{-}paste} + R_{substrate}$$
$$+ R_{TIM} + R_{heatsink}.\qquad(1.8)$$

If low-conductivity materials were used, that is, silicon die attach, Ag solder paste, and grease as TIM, Equation (1.8) can be simplified as

$$R_{tot} = R_{die} + R_{slug} + R_{substrate} + R_{heatsink}.\qquad(1.9)$$

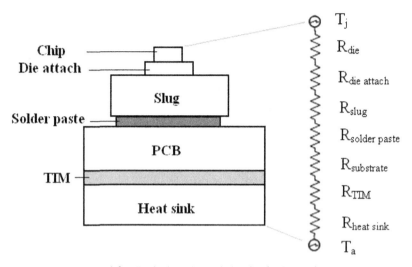

Figure 1.2   Equivalent thermal circuit of LED package.

By assuming a uniform heat flux on the top of the chip, the thermal resistance of the chip can be described with a 1D thermal resistance model for heat diffusion:

$$R_{\text{die}} = \frac{t_{\text{die}}}{k_{\text{die}} A_{\text{die}}}, \tag{1.10}$$

where t is the thickness, k is the thermal conductivity, and A is the cross-sectional area. Similarly, thermal resistance of the slug can be expressed as

$$R_{\text{slug}} = \frac{t_{\text{slug}}}{k_{\text{slug}} A_{\text{slug}}}. \tag{1.11}$$

The structure model of the LED package with multiple chips is shown in Figure 1.3. The LED chip is with an area of $c \times d$ with a center coordinate of (X_i, Y_i), mounted on the heat spreader measured in $a \times b$. In general, the heat spreader was insulated on all surfaces, except at the bottom which was left exposed to ambient.

A general analytical solution based on the separation variable method can be employed for temperature estimation on rectangular flux channels. The solution for the temperature distribution on the heat sink, which calculates the temperature of multiple chips LED package, was stated as (Cheng et al., 2010)

$$T(x, y, z) - T_{\text{f}} = \Delta T = \sum_{i=1}^{N} \theta_i(x, y, z), \tag{1.12}$$

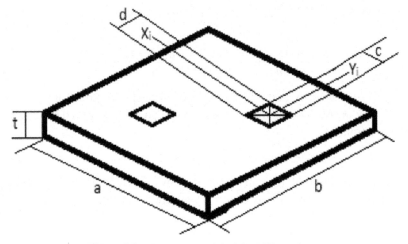

Figure 1.3 Structure model of the LED package.

where T_f is the ambient temperature, $T(x, y, z)$ is the temperature of the LED at the coordinate (x, y, z), N is the number of the chip, and θ_i measures the total temperature excess of the module. The following equation is to express the surface temperature distribution of the LED module at $z = 0$:

$$\theta_i (x, y, 0) = A_0^i + \sum_{m=1}^{\infty} A_m^i \cos (\lambda x) + \sum_{n=1}^{\infty} A_n^i \cos (\delta x)$$
$$+ \sum_{m=1}^{\infty} \sum_{n=1}^{\infty} A_{mn}^i \cos (\lambda x) \cos (\delta x). \tag{1.13}$$

Several Fourier coefficients, that is, A_m, A_n, and A_{mn}, must be determined in order to obtain the temperature excess. A_0 is the value for the coefficient in uniform flow and is calculated by

$$A_0 = \frac{Q}{ab} \left(\frac{t}{k} + \frac{1}{\gamma} \right), \tag{1.14}$$

where γ indicates the heat transfer coefficient of the LED to the base, k is the thermal conductivity of the heat spreader, and Q is the dissipation power. The Fourier coefficients can be determined by

$$A_m^i = \frac{2Q \left[\sin (\lambda_m (2X_i + c_i)/2) - \sin (\lambda_m (2X_i + c_i)/2) \right]}{abc_i k \lambda_m^2 \varphi (\lambda_m)}, \tag{1.15}$$

$$A_n^i = \frac{2Q\left[\sin\left(\delta_n\left(2Y_i + d_i\right)/2\right) - \sin\left(\delta_n\left(2Y_i - d_i\right)/2\right)\right]}{abd_i k \delta_n^2 \varphi\left(\delta_n\right)}, \text{ and} \quad (1.16)$$

$$A_{mn}^i = \frac{16Q\cos\left(\lambda_m X_i\right)\sin\left(\lambda_n c_i/2\right)\cos\left(\delta_n y_i\right)\sin\left(\delta_n d_i/2\right)}{abc_i d_i k \beta_{m,n} \lambda_m \delta_n \varphi\left(\beta_{m,n}\right)}, \quad (1.17)$$

where $\lambda = m\pi/a$, $\delta = n\pi/b$, and $\beta = \left(\lambda^2 + \delta^2\right)^{0.5}$. For the multiple chip LED package, $N \times N$ array, the total thermal resistance can be determined by using the following equation:

$$R_{\text{tot}} = R_{\text{die}} + R_{\text{slug}} + R_{\text{substrate}} + N \times R_{\text{heatsink}}. \quad (1.18)$$

The last term in Equation (1.18) can be obtained by considering the innermost LED which gives a maximum junction temperature:

$$N \times R_{\text{heatsink}} = \frac{T\left(x, y, 0\right)}{NQ}. \quad (1.19)$$

1.4 Simulation

Simulation study of the LED package can be made with the help of the professional thermal analysis software. The software Flotherm can be utilized to perform the thermal modeling work. With the embedded CFD solver, Flotherm is able to solve the Navier-Stokes equations for mass, momentum, and energy conservation with the finite volume technique (Panton, 1996). The equation for mass can be expressed as

$$\left[\frac{\partial \rho}{\partial t} + v \cdot \nabla \rho\right] = -\rho \nabla \cdot v, \quad (1.20)$$

where ρ is the density, t is the time, and v is the velocity vector. Momentum is another vector measurement which is in the same direction as velocity and is calculated as

$$\rho\left[\frac{\partial v}{\partial t} + v \cdot \nabla v\right] = -\nabla p + \rho g - 0.67 \cdot \nabla\left(\mu\nabla \cdot v\right) + 2\nabla \cdot \mu S, \quad (1.21)$$

where p is the pressure, g is the gravity, μ is the viscosity, and S is the strain rate tensor. The thermal energy is presented as follows:

$$\rho c_{\text{P}}\left[\frac{\partial T}{\partial t} + v \cdot \nabla T\right] = \nabla \cdot \left(k\nabla T\right) - 0.67 \cdot \mu\left(\nabla \cdot v\right)^2 + 2\left(\mu S : S\right)$$

$$+ \beta T \cdot \left[\frac{\partial p}{\partial t} + \left(v \cdot \nabla\right) p\right], \quad (1.22)$$

where T is the absolute temperature, k is the thermal conductivity, and β is the expansiveness. A number of simulation softwares could be used to perform the thermal analysis, for example, Star CCM+ and ANSYS. STAR-CCM+ performs thermal modeling of electric machines and calculates the temperature transient of the system. STAR-CCM+ is able to solve engineering problems involving fluid flow, conjugate heat transfer, and solid stress. On the other hand, ANSYS allows the design and optimization of new equipment and to troubleshoot existing installations. Furthermore, it contains modeling capabilities to model flow, turbulence, heat transfer, and reactions for real-case industrial applications.

1.5 General Results on Thermal Field Analysis

The experimental evaluation of thermal study can be implemented by the use of a thermal transient tester. Thermal transient tester enables the measurement of the thermal characteristics of packaged semiconductor devices. Figure 1.4 illustrates the setup and connections of the system. The LED module was placed on a cooling plate where temperature was regulated by the thermostat.

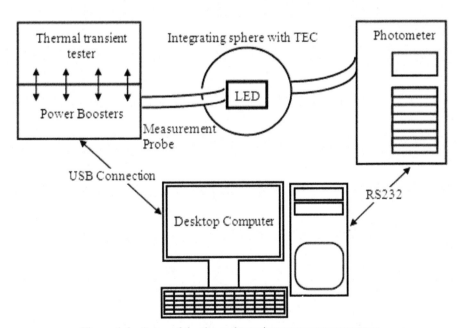

Figure 1.4   Setup of the thermal transient measurement system.

Thermoelectric cooler (TEC) provided a constant temperature (25°C) environment for the LED module which was assumed to have infinite thermal capacitance. The power booster interface with the LED module through the measurement probes indicated the thermal characteristic of the LED module. In addition to the experimental evaluation, the heat distribution on the influence of junction temperature was simulated. The analysis of both experimental and simulation results enables us to make some general predictions regarding the thermal performance of the LED package in terms of different parameters.

1.5.1 In Terms of *K* Factor

The *K* factor, which determines the ratio between forward voltage drop on the LED with specific temperature at a given operating current, was calculated as

$$K \text{ factor} = \frac{\Delta V}{\Delta T}. \tag{1.23}$$

Figure 1.5 shows the *K* factor of three LED modules with single-chip, double-chip, and four-chip arrangement. The steady-state forward voltage of the LED was measured at a temperature ranging between 25 and 45°C in an ascending order. The linear relationship can be obtained between junction temperature and forward voltage. As in Figure 1.5, a nearly constant value of *K* factor was

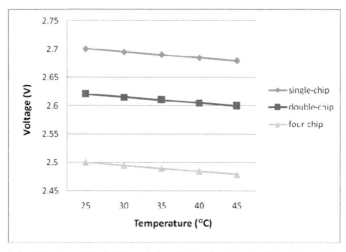

Figure 1.5   *K* factor of the LED module with single-chip, double-chip, and four-chip.

calculated as −1.05 mV/°C regardless of the number of chips on the module. It implies a 1°C increase of junction temperature inducing 1.05 mV decreases in forward voltage.

1.5.2 In Terms of Cooling Curve

The cooling curves for LED packages with single-chip, double-chip, and four-chip are summarized in Figure 1.6. It is apparent that the junction temperature increases with the number of chips, or to be precise, with the input power. Figure 1.7 shows the cooling curve of LED packages with Al substrates under different interface boundary conditions. Obviously, an AlN/Al stack thin film deposited on the Al substrate successfully reduces the junction temperature in comparison with the AlN thin film, both with tp and without tp.

1.5.3 In Terms of Structure Functions

The differential structure function as shown in Figure 1.8 exhibits a total thermal resistance of 2.23 K/W. As discussed previously in Section 1.3.2, the total thermal resistance can be expressed as

$$R_{\text{tot}} = R_1 + R_2 + R_3 + R_4, \tag{1.24}$$

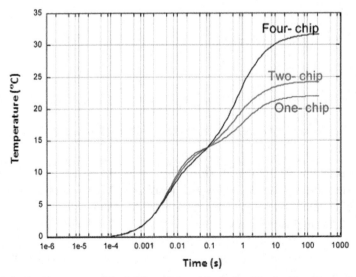

Figure 1.6 Cooling curve of the LED module with single-chip, double-chip, and four-chip (Kim and Shin, 2007).

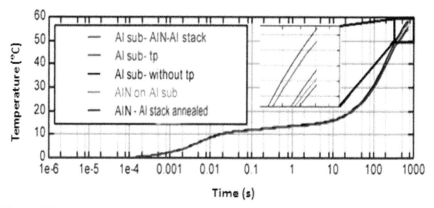

Figure 1.7   Cooling curve of the LED module with different interface boundary conditions at 350 mA (Subramani and Devarajan, 2014a).

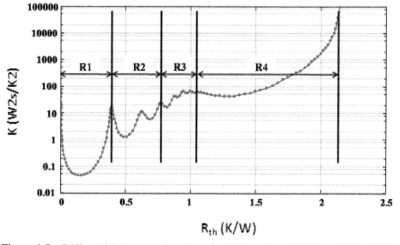

Figure 1.8   Differential structure function of a LED package (Choi and Shin, 2012).

where R_1 denotes the thermal path from chip to PCB, R_2 denotes the path from the PCB to the Al case, R_3 is the path within the Al case, and R_4 is from the heat sink to the ambient. It is desirable to reduce thermal resistance for ensuring good thermal performance.

The cumulative structure function is shown in Figure 1.9, where the vertical axis represents the thermal conductance and the horizontal axis represents the thermal resistance. The figure shows that the Al substrate with a BN coating has a better performance on thermal resistance.

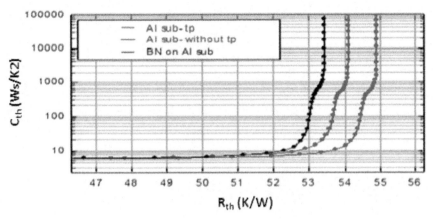

Figure 1.9   Cumulative structure function of the LED module with different interface boundary conditions at 350 mA (Subramani and Devarajan, 2014b).

1.5.4 In Terms of Estimated Junction Temperature

The thermal simulation results for different chips are shown in Figure 1.10. It is clear that the junction temperature appears as the hottest spot and to be distributed to the ambient of 35°C. The reliability and the stability of the LED package can be maintained as long as the junction temperature remains below the critical transition temperature of 120–130°C.

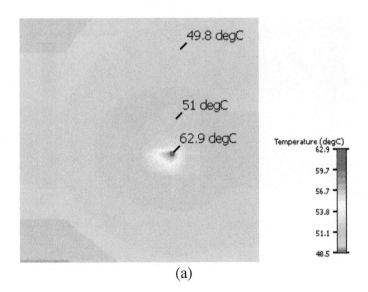

(a)

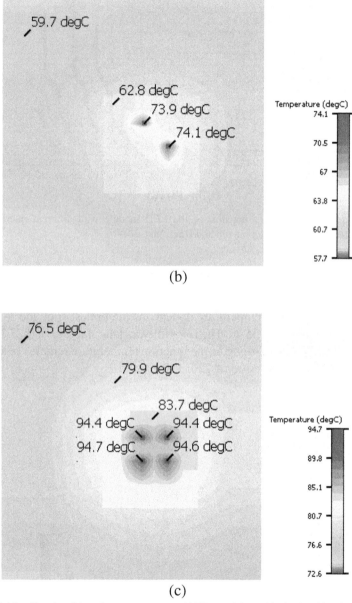

Figure 1.10   Simulated junction temperature of LED modules with single-chip, double-chip, and four-chip.

References

[1] Anderson KR., Devost M., Pakdee W. and Krishnamoorthy N. (2013). STAR CCM+ CFD simulations of enhanced heat transfer in high-power density electronics using forced air heat exchanger and pumped fluid loop cold plate fabricated from high thermal conductivity materials. *Journal of Electronics Cooling and Thermal Control*, 3: 144–154.

[2] Ashry M. (2015) Augmenting the heat sink for better heat dissipation. *Circuits and Systems*, 6, 21–29. doi: 10.4236/cs.2015.62003

[3] Cheng HH., Huang D-S. and Lin M-T. (2012). Heat dissipation design and analysis of high power LED array using the finite element method. *Microelectronics Reliability*, 52: 905–911.

[4] Cheng T., Luo X., Huang S. and Liu S. (2010). Thermal analysis and optimization of multiple LED packaging based on a general analytical solution. *International Journal of Thermal Sciences*, 49(1): 196–201.

[5] Choi JH. and Shin MW. (2012). Thermal investigation of LED lighting module. Microelectronics Reliability, 52(5): 830–835.

[6] Chuang SL., Ishibashi A., Kijima S., Nakayama N., Ukita M. and Taniguchi S. (1997). Kinetic model for degradation of light emitting diodes. *IEEE Journal of Quantum Electronics*, 33: 970–979.

[7] Darwish AM., Bayba AJ., and Hung HA. (2004). Thermal resistance calculation of AlGaN–GaN devices. *IEEE Transactions on Microwave Theory and Techniques*, 52: 2611–2620.

[8] Farkas G., vanVoorstVader Q., Poppe A., and Bognár G. (2005). Thermal investigation of high power optical devices by transient testing. *IEEE Transactions on Components and Packaging Technologies*, 28(1): 45–50.

[9] Ha M. and Graham S. (2012). Development of a thermal resistance model for chip-on-board packaging of high power LED arrays. *Microelectronics Reliability*, 52: 836–844.

[10] Kim L. and Shin MW. (2007). Thermal resistance measurement of LED package with multichips. *IEEE Transactions on Components and Packaging Technologies*, 30(4): 632–636.

[11] Li J., Wang D. and Peterson GP. (2009). Development of a robust miniature loop heat pipe for high power chip cooling. In *ASME 2009 Second International Conference on Micro/Nanoscale Heat and Mass Transfer* (pp. 347–354). American Society of Mechanical Engineers, New Year.

[12] Liu DJ., Yang DG., You Z. and Hou FZ. (2011). Spreading Resistance Analysis of LED by Structure Function. *Advanced Materials Research*, 199–200: 1501–1504.

[13] Luo X., Cheng T., Xiong W., Gan Z. and Liu S. (2007). Thermal analysis of an 80 W light-emitting diode street lamp. *IET Optoelectronics*, 1: 191–196.

[14] Ma Z. and Zheng X. (2007). Thermal resistance calculation method of high-power LEDs. Science Technology and Engineering, 7: 1671–1819.

[15] Meneghini M., Tazzoli A., Mura G., Meneghesso G. and Zanoni E. (2010). A review on the physical mechanisms that limit the reliability of GaN-based LEDs. *IEEE Transactions on Electron Devices*, 57: 108–118.

[16] Narasimhan S., Bar-Cohen A. and Nair R. (2003). Flow and pressure field characteristics in the porous block compact modeling of parallel plate heat sinks. *IEEE Transactions on Components and Packaging Technologies*, 26(1): 147–157.

[17] Pan K., Lin H., Guo Y., Wei N., Lu T. and Zhou B. (2014). Study on the thermal resistance of multi-chip module high power led packaging heat dissipation system. *Sensors and Transducers*, 180(10): 72–79.

[18] Panton R. (1996). *Incompressible flow*. New York: John Wiley, pp. 129–46.

[19] Perpiñà X., Werkhoven R., Jakovenko J., Formánek J., Vellvehí M., Jordà X., et al. (2012). Design for reliability of solid state lighting systems. *Microelectronics Reliability*, 52: 2294–2300.

[20] Poppe A., Horváth G., Nagy G., Rencz M. and Székely V. (2008). Electro-thermal and logi-thermal simulators aimed at the temperature-aware design of complex integrated circuits. *In Twenty-fourth Annual IEEE Semiconductor Thermal Measurement and Management Symposium, 2008.* Semi-Therm 2008 (pp. 68–76).

[21] Steranka FM., Bhat J., Collins D., Cook L., Craford MG., Fletcher R., et al. (2002). High power LEDs – Technology status and market applications. *Physica Status Solidi* (a), 194: 380–388.

[22] Subramani S. and Devarajan M. (2014a). Influence of AlN Thin Film as Thermal Interface Material on Thermal and Optical Properties of High-Power LED. *IEEE Transactions on Device and Materials Reliability*, 14(1): 30–34.

[23] Subramani S. and Devarajan M. (2014b). Thermal transient analysis of high-power green LED fixed on BN coated al substrates as heatsink. *IEEE Transactions on Electron Devices*, 61(9): 3213–3216.

[24] Székely V. (1991). On the presentation of infinite-length distributed RC one-ports. *IEEE Transactions on Circuits and Systems*, 38(7): 711–719.

[25] Tsai MY., Chen CH. and Kang CS. (2012). Thermal measurements and analyses of low-cost high-power LED packages and their modules. *Microelectronics Reliability*, 52(5): 845–854.

[26] Williams EL., Haavisto K., Li J. and Jabbour GE. (2007). Excimer-based white phosphorescent organic light-emitting diodes with nearly 100% internal quantum efficiency. *Advanced Materials*, 19: 197–202.

[27] Yamamoto H., Kimura T., Matsumoto S. and Suyama S. (2010). Viewing-zone control of light-emitting diode panel for stereoscopic display and multiple viewing distances. *Journal of Display Technology*, 6(9): 359–366.

[28] Yang L., Jang, S., Hwang, W. and Shin M. (2007). Thermal analysis of high power GaN-based LEDs with ceramic package. *Thermochimica Acta*, 455: 95–99.

2

Modeling and Simulation of Ge/Si-Nanodisk Array for QD-based IB Solar Cells

Ming-Yi Lee[1], Yi-Chia Tsai[1], Yiming Li[1] and Seiji Samukawa[2]

[1]Parallel and Scientific Computing Laboratory, Institute of Communications Engineering, Department of Electrical and Computer Engineering, National Chiao Tung University, Hsinchu 300, Taiwan
[2]Institute of Fluid Science and WPI-AIMR, Tohoku University, Sendai 980-8577, Japan

Abstract

A computationally efficient finite element simulation is performed to calculate the miniband structure and DoSs for the well-ordered Ge/Si-nanodisk array. The semiconductor nanostructures are fabricated by using self-assemble bio-template and damage-free NBE technique. Within the envelop-function framework, our model surmounts theoretical approximations of the multi-dimensional Kronig–Penney method and accurately calculates the energy dispersion relationship. The miniband formation works as the intermediate band within the bandgap of bulk silicon band. Effects of the interdot space, the radius and thickness of the Ge/Si-nanodisk on the miniband structure, and conversion efficiency of the solar cell (SC) are discussed. The findings of this study provide a guideline for 3D QDs IBSC design.

Keywords: Ge/Si-Nanodisk array, IBSC, FEM.

2.1 Introduction

Among the next-generation high-efficiency solar cell (SC) technologies, a promising candidate is the use of QDs. The QDs work as the IB in the SC and dramatically increase the efficiency of converting sunlight into energy

Green Photonics and Smart Photonics, 25–46.

because of their ability to absorb sub-bandgap photons through the two-photon transition from the VB to the CB. The detailed balance-limiting efficiency of 63% attracted great interest in the theoretical and experimental researches [1–7]. When uniform QDs are closely packed as the superlattice, the wavefunction of each QD couples with neighboring QDs to broaden the discrete quantum levels to form finite-width minibands. Miniband structure is a key parameter which determines two-photon transition and photo-generated carrier transport for QDs SC application. Researchers [8, 9] extended the analytic Kronig-Penney method to describe the 3D QDs superlattice and offer significant information for QDs SC design with approximations of the independent periodic potential in the quantum cuboid system. Given the rapid development of nanotechnologies and device processes, it would be better if the more accurate simulation method without any constraints on QDs' structure is developed to instruct realistic QDs fabrication and design. For example, NBE, a promising top-down nanotechnology, when combined with bio-template, provides greater flexibility in engineering quantum structures such as independently adjustable diameter, thickness, interdot space, incline angle, matrix materials, and so on [10, 11]. The FEM was developed in our recent study [12] to simulate a realistic Si/SiC array in the whole real space. However, the numerical simulation in real space is computationally expensive due to the large memory size for mesh storage and is time-consuming for large-scale eigenvalue problem.

In this chapter, a new 3D FEM model is proposed to more efficiently simulate the complex and realistic physical properties, such as miniband structure and DoSs, in a periodic QDs array. This method surmounts theoretical approximations of the multidimensional Kronig-Penney method and reveals some significant information for in-plane 3D QDs array design.

2.2 Fabrication of Ge/Si-Nanodisk

Compared to kinetic-driven bottom-up nanotechnologies, top–down nanotechnologies are more attractive for fabrication of QDs superlattices. However, the traditional top–down process with the lithography process is hardly implanted into quantum size fabrication. The first difficulty is the physical limitation of the conventional photolithographic mask, which is larger than the semiconductor exciton Bohr radius (Si is 4.9 nm). The second difficulty is unavoidably damage induced by the common plasma etching, which act as recombination centers to degrade optical and electrical properties. By combining the self-assemble bio-template and damage-free NBE, a top–down

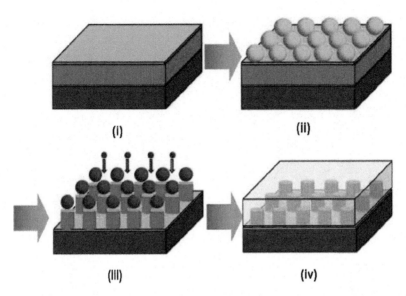

(i)   (ii)

(III)   (iv)

Figure 2.1 An advanced top–down nano-fabrication technology by combining the self-assemble bio-template and damage-free NBE [6]: (i) Ge/Si stacked layers deposition; (ii) a two-dimensional array ferritin self-assembly and protein shell removal; (iii) damage-free NBE; (iv) matrix regrowth.

process illustrated in Figure 2.1 was developed to fabricate sub-10 nm uniform and well-aligned type-II Si/Ge QDs superlattice [13, 14]. At first, germanium and silicon are alternately deposited on substrates. The in situ measurement is used to precisely control thickness. Next, ferritin molecules self-assemble to form a 2D array with the spin-coating method. Then, high-temperature annealing removes protein shells so that 2D of metal oxide cores is left as the etching mask. The oxidation surface is isotropically etched by the NF_3 and then is anisotropically etched by the Cl_2 neutral beam to form QDs. Finally, matrix silicon is deposited to finish the high-periodical superlattice. The etched Ge/Si nanopillars before regrowth matrix in Figure 2.2 exhibit a good uniformity and alignment.

This nanofabrication technique can control the QD thickness by the deposition thickness and its diameter by the bio-template, which brings higher flexibility on engineering quantum levels. With a stable DNA duplicate, the QD superlattice has high uniformity and quasicrystalline in-plane alignment. At the same time, the ideal vertical alignment is achieved with one-step etching on stacked Ge/Si layers. Moreover, the regrowth process is independent with the etching process, which indicates various possible matrix materials.

Figure 2.2 SEM image of Ge/Si nanopillars with a uniform density and regular arrangement by using ferritin iron cores as an etching mask before regrowth matrix for advanced top–down nano-fabrication technology as Figure 2.1.

2.3 The Computational Model

2.3.1 Calculation of Electronic Band Structures

Within the one-band envelope-function theory, the electronic structure can be described by the Schrödinger equation with the effective mass approximation as

$$\nabla \left[-\frac{\hbar}{2m^*} \nabla \psi(r) \right] + V(r)\psi(r) = E\psi(r), \tag{2.1}$$

where $\hbar, m^*, V(r), E$, and $\psi(r)$ are the reduced Planck'sconstant, the effective mass, the position-dependent potential energy, quantum energy levels, and the envelope function respectively. For a periodic superlattice, such as in-plane 3D QDs array described by two primitive vectors a_1 and a_2 in real space, the superlattice is under the presence of a periodic potential energy $V(r+R) = V(r)$, where $R = n_1 a_1 + n_2 a_2$ is the lattice vector with integers n_1 and n_2. The envelope function will satisfy Bloch theorem and have the Bloch wave form.

$$\psi(r) = \exp\left(ik \cdot r \right) u_k(r), \tag{2.2}$$

where k is the lattice wave vector in irreducible Brillouin zone (IBZ). With the Bloch wave form Equation (2.2), Equation (2.1) becomes

$$\nabla \left[-\frac{\hbar}{2m^*} \nabla u_k(r) \right] - \frac{i\hbar}{2m^*} k \cdot \nabla u_k(r)$$

$$+ \left[V(r) + \frac{\hbar k^2}{2m^*} \right] u_k(r) = E_{n,k} u_k(r), \qquad (2.3)$$

n is the quantum number and function $u_k(r)$ follows the periodic boundary condition

$$u_k(r + R) = u_k(r). \qquad (2.4)$$

Figure 2.3 shows the simulation flow to solve the bloch function $u_k(r)$ and eigenvalue $E_{n,k}$ for a highly periodic superlattice. First, a unit cell formed by the primitive vectors is defined. Then, based on the symmetry of the square superlattice, the k-points space is defined in a triangular IBZ. Finally, to get band structure $E_{n,k}$ and Bloch function $u_k(r)$, Equation (2.3) is discretized within a unit cell in real space and solved by a FEM solver with a boundary condition of Equation (2.4) for each sampling k-point in IBZ. Comparing to the whole real-space simulation [12], the required memory size for mesh storage and the computation time for the eigenvalue problem

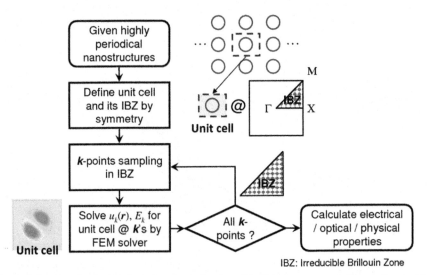

Figure 2.3 The simulation flow chart for an in-plane 3D QDs square superlattice and material parameters used for simulation [4].

of the Hamiltonian matrix can be reduced by hundreds at least for QDs in 10 times 10 arrays because the real-space simulation is executed only within a unit cell. Further, the proposed method does not have constraints on the geometry and structure of periodic QDs superlattice.

2.3.2 Calculation of DoSs

Based on the energy distribution in IBZ, the DoS can be calculated. The DoS, the number of allowed carrier states per unit area per unit energy, is a key parameter for electronic and optical applications of the semiconductor system. By way of definition of the DoS

$$g(E) = \frac{2}{(2\pi)^2} \int_{BZ} \frac{dl_E}{|\nabla_k E_{n,k}|} = \frac{2}{(2\pi)^2} \sum_{n,l} \frac{l_n(E, K_l)}{|\nabla_k E_n(k_l)|}, \quad (2.5)$$

where l_E denotes the line of constant energy E in the 2D k-space. The factor 2 accounts for the twofold electron spin degeneracy. The integral is carried out numerically using an improved triangle method [15, 16] by dividing the IBZ into a large number of small triangular cells, as in Figure 2.4. Each triangular cell l with area A_l in k-space contributes to DoSs $g_{n,l}(E)$ by

$$g_{n,l}(E) = \begin{cases} 0 & \text{if } E < E_1 \\ \frac{A_l}{A_G} \frac{2(E-E_1)}{(E_2-E_1)(E_3-E_1)} & \text{if } E_1 < E < E_2, \\ \frac{A_l}{A_G} \frac{2(E_3-E)}{(E_3-E_1)(E_3-E_2)} & \text{if } E_2 < E < E_3 \\ 0 & \text{if } E_3 < E \end{cases} \quad (2.6)$$

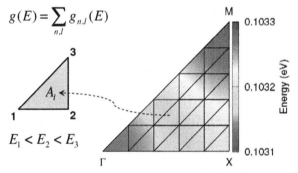

Figure 2.4   The IBZ is divided into a large number of small triangular cells. Each triangular cell l with area A_l in k-space contributes to DoSs $g_{n,l}(E)$ by Equation (2.6). The number of corner for each cell is sorting based on energy by $E_1 < E_2 < E_3$.

where E_i for $i = 1$ to 3 is the energy of triangular cell corner sorting in the order of $E_1 < E_2 < E_3$, and A_G is the area of IBZ.

2.3.3 IBSC Model

To investigate the effect of miniband on the ultimate conversion efficiency of the SC, the following assumptions are used in the QDs IBSC model [1, 2, 17]:

1. The SC is thick enough, such as multilayers of in-plane QDs array, to have full absorption of the photons with energy larger than subbandgaps.
2. Non-radiative transitions recombination between bands is forbidden.
3. The IB is electrically isolated so that no carrier can be extracted from the IB except that of photon-induced transition.
4. The quasi Fermi levels are constant through the SC so that photon-induced carriers contribute all the current except of carrier loss due to the non-radiative recombination.

As illustrated in Figure 2.5, with the detailed balance arguments of Luque theory [1, 17] for IBSC, the electron current density is given by

$$\begin{aligned} J &= J_{\text{CV}} + J_{\text{IB}} \\ &= e\left(G_{\text{CV}} + G_{\text{CI}} - R_{\text{CV}} - R_{\text{CI}}\right), \end{aligned} \tag{2.7}$$

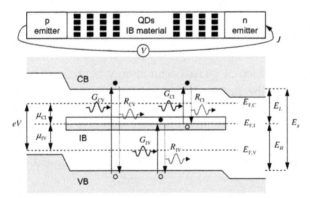

Figure 2.5 The solar cell consists of QDs as IB materials and p–/n–junction as emitter at both ends of the terminals. The band diagram shows the bandgap E_g due to bulk material, the subbandgap E_L and E_H due to the IB formed by QDs, as well as the quasi-Fermi levels $E_{F,C}$, $E_{F,V}$, and $E_{F,I}$ for the CB, VB and IB respectively with biased at voltage V. The generation rate $G_{I,J}$ represents photon absorption between band I and band J while the recombination rate R_{IJ}, represents photon emission between band I and band J, which depends on the difference of quasi-Fermi levels between bands μ_{IJ}.

where $J_{CV} = e\,(G_{CV} - R_{CV})$ are photocurrent density induced from the bulk matrix materials [1] and $J_{IB} = e\,(G_{CI} - R_{CI})$ are photocurrent density induced from the intermediate bands created by QDs array [17].

The generated electrons per unit of area and time due to photon fluxes absorbed by the cell through transition of band I and band J, G_{IJ} is given by

$$G_{IJ} = \chi \Omega_s \int f(E, T_s, 0)\mathrm{d}E + (\pi - \chi \Omega_s) \int f(E, T_c, 0)\mathrm{d}E, \qquad (2.8)$$

where Ω_s is a geometrical factor determined by the sunlight angle $\Theta \approx 0.53°$ as $\pi \sin^2(\Theta/2)$, χ is the concentration used in a loss-less concentrator so that the maximum of $\chi \Omega_s$ is π, T_s, and T_c are the temperatures of sun and SC, respectively, and f is the generalized Planck formula

$$f(E, T, \mu) = \frac{2}{h^3 c^2} \frac{E^2}{\exp\left((E - \mu)/kT\right) - 1}. \qquad (2.9)$$

With Equation (2.9), $f(E, T_s, 0)$ describes the photon flux distribution from the sun when assumed as a black body at temperature T_s, while $f(E, T_c, 0)$ are the thermal photon flux from the cell at ambient temperature T_c.

On the other hand, associated with the generation process described by Equation (2.8), R_{CV} and R_{CI} are the radiative recombination of electrons per unit of area and time to emit photon fluxes by the cell through transition of CB to VB and CB to IB, respectively [18]. The recombination rate is formulated as

$$R_{IJ} = \pi \int f(E, T_c, \mu_{IJ})\mathrm{d}E, \qquad (2.10)$$

where μ_{IJ} is the difference of quasi-Fermi energy between miniband I and miniband J [1, 19]. In the steady state, no current is extracted from the IBs and flows between the IBs under assumption of constant quasi-Fermi energy so that

$$G_{CI} - R_{CI}(\mu_{CI}) = G_{IV} - R_{IV}(\mu_{IV}). \qquad (2.11)$$

For an applied external voltage V, the difference of quasi-Fermi energy is also determined by

$$V = \mu_{CV} = \mu_{CI} + \mu_{IV}. \qquad (2.12)$$

Using Equations (2.11) and (2.12), we can get the two unknown variables μ_{CI} and μ_{IV} for a given external voltage V and then obtain the current–voltage (J–V) characteristics by Equation (2.7) as well as the conversion efficiency.

To consider the density of QDs, Equation (2.7) is further modified as

$$J = J_{CV}(\mu_{CV}) + \nu J_{IB}(\mu_{CI}, \mu_{IV}), \qquad (2.13)$$

where ν is the volume ratio of QDs to bulk matrix that contributes to J_{IB} from the IB localized in QDs since the IB comes from the confinement effect of QDs in bulk materials and its wavefunction is localized within QDs.

2.4 1D Superlattice

For a 1D superlattice with periodic square potential energy as shown in Figure 2.6, the band structure can be calculated by the analytical Kronig-Penney method [20]:

$$\cos(kR) = \cos(k_a a)\cos(k_b b)$$
$$-\frac{1}{2}\left(\frac{k_b m_a^*}{k_a m_b^*} + \frac{k_a m_b^*}{k_b m_a^*}\right)\sin(k_a a)\sin(k_b b), \qquad (2.14)$$

where R is sum of the well thickness a and the barrier thickness b, m_i is the effective mass in the well or barrier, V_i is the potential energy in the well or barrier, and $k_i = \sqrt{2m_i^*(E - V_i)}$.

To compare the analytical Kronig-Penney method with the proposed method, the parameters used for simulation are listed in Table 2.1. The results of our FEM method are shown in the form of solid symbols in Figure 2.7. From this figure, we can find that the calculated energy levels (symbols) match well with the energy in the same wave vector from the analytic model Equation (2.14) (lines in Figure 2.7). Thus, the proposed method accurately simulates the superlattice system by a unit cell with limited sampling k-points in IBZ.

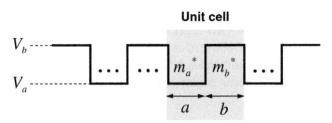

Figure 2.6   Plot of 1D superlattice with periodic square potential and used value of parameters are in Table 2.1.

Table 2.1   Material parameters used in the simulation

Parameter	a (nm)	b	V_a (eV)	V_b (eV)	M_a^*	M_b^*
Value	4	4 nm	0	0 eV	0.28 m_e	0.49 m_e

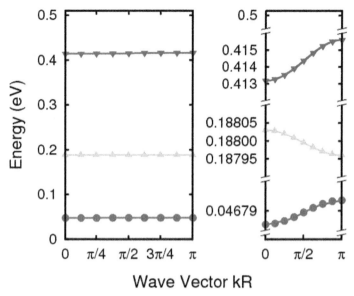

Figure 2.7   The band dispersion relation for 1D periodic square-potential superlattice with parameters in Table 2.1 (symbols: simulation, lines: analytical Kronig-Penney method). The right figure with broken segments in different scale to reveal the agreement between simulation and analytical Kronig-Penney method.

2.5 In-Plane Ge/Si QDs Superlattice

Although the silicon-based SC has more than 85% market share due to its abundant resource of material silicon, the conversion efficiency is up to the Shockley-Queisser limitation of 29.43% [21]. Based on a top-to-down nanotechnology by neutral beam etching combined with a bio-template [6], a highly uniform 3D type-II Ge QDs array in the bulk Si matrix indicates the possibility to achieve 44% of conversion efficiency [13, 14]. The proposed method is used to study the miniband structures of the idealistic Ge QDs array in the bulk Si matrix. The material parameters used in the simulation are shown in Table 2.2.

Table 2.2   Material parameters used in the simulation [4]

	Ge QDs		Si Matrix
Effective mass (m_e)	0.28		0.49
Bandgap (eV)	0.66		1.12
Band barrier (eV)		0.51	

2.5.1 Electronic Band Structure

As in Figure 2.8, an in-plane 3D QDs array is interpreted by a 2D square superlattice that can be described by a cuboid unit cell in real space with a triangular IBZ in k-space. A FEM solver is used to discretize the Schrödinger equation Equation (2.3) in the real-space unit cell for each sampling k-points of IBZ.

Figure 2.9 shows the calculated band dispersion relation for QD in dimension of the radius 2 nm and the thickness 4 nm and the interdot space 2.3 nm [6]. There are totally six bounded states (minibands) with energy lower than band barrier 0.51 eV. Because the wavefunction of each QD couples with neighboring QDs, the energy structures are not in discrete levels but in finite-bandwidth minibands. However, the bandwidth of bounded minibands is small ($<< kT \sim 0.025$ eV), which implies the weak interaction between QDs under the interdot space 2.3 nm.

Figure 2.10 shows the isosurface of wave function within Ge QD at Γ-point for lowest four bounded-states. The ground state has an ellipsoid surface without nodes as the expected s-orbit so that its energy distribution is isotropic in k-space, as in Figure 2.11a. The first excited state shows a p-orbit along z-direction so that its energy distribution is isotropic in 2D k-space also, as in Figure 2.11b. The second and third excited states show two-fold degenerate p-orbits due to the symmetry of the x–y plane in QDs superlattice

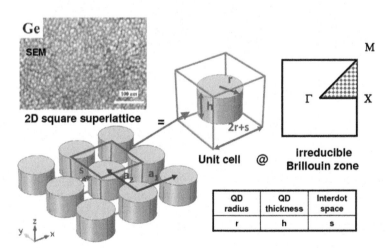

Figure 2.8 Schematic illustration for a realistic in-plane 3D QDs array [14] by a 2D square superlattice that is interpreted (=) as a unit cell formed by two primitive vectors a_1 and b_1 with k-points restricted (@) in IBZ.

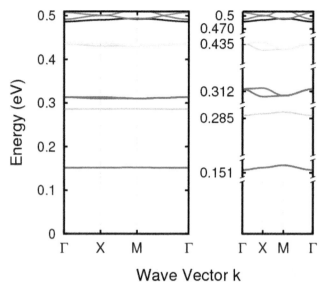

Figure 2.9 The band dispersion relation for 3D in-plane Ge/Si QDs superlattice with QD's radius 2 nm, thickness 4 nm and interdot space 2.3 nm. The right figure breaks energy axis into several segments in different scale to highlight the profile of each quantum level. There are totally SIX minibands with energy lower than band barrier 0.51 eV.

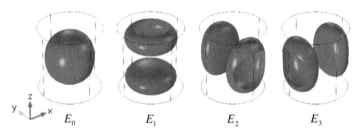

Figure 2.10 The isosurface of wave function within QD in dimension of radius 2 nm, thickness 4 nm and interdot space 2.3 nm at Γ-point of IBZ for lowest four bounded-states: s-orbit E_0, non-degenerate p-orbit E_1, and degenerate p-orbits E_2 and E_3.

so much so that its energy distribution is constant in k_y or k_x direction in IBZ. This also explains the degeneracy of energy dispersion for the second and third excited states in Figure 2.9 along the symmetry points M and Γ. The energy distributions and wave functions are closely related and reveal the symmetry of the QDs superlattice.

In addition to the band structure, the DoS can also be calculated by the energy distribution in IBZ. In the next section, the dependence on interdot

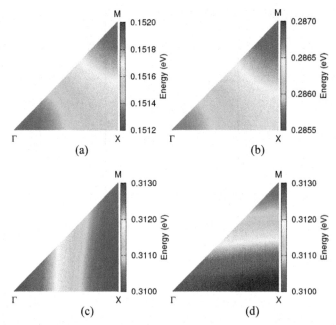

Figure 2.11 The energy distribution in IBZ for QDs superlattice in dimension of radius 2 nm, thickness 4 nm, and interdot space 2.3 nm: (a) ground bounded-band $E_{0,k}$, (b) first excited bounded-band $E_{1,k}$, (c) second excited bounded-band $E_{2,k}$, and (d) third excited bounded-band $E_{3,k}$.

space, thickness and radius of QDs superlattice is individually investigated by the DoS for providing a guideline for miniband design.

2.5.2 Density of States

Based on the energy distribution in IBZ, the DoSs can be calculated by Equation (2.5). For the top-to-down nanotechnology of NBE combined with a bio-template, the two independently controllable structure parameters, radius (interdot space), and thickness of QDs superlattice, bring higher flexibility in miniband design [22].

Figure 2.12 presents the results for QDs superlattice with varied interdot space from 3.3 to 0.3 nm, radius 2 nm, and thickness 4 nm. As the interdot space decreases, QDs interaction between discrete levels increases and the bandwidth of the miniband also increases, and the miniband mixing phenomenon occurs especially for higher excited states. This phenomenon indicates the strong dependence of interdot space on the coupling strength.

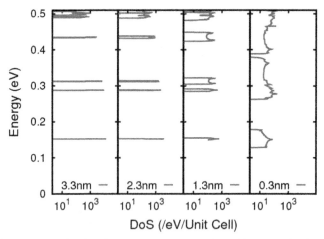

Figure 2.12 The DoSs for Ge/Si QDs square superlattice with QD radius 2 nm, thickness 4 nm, and varied interdot space from 3.3 to 0.3 nm.

Because the higher excited states are mixed and become continuous energy levels to Si barrier, the effective bandgap of bulk Si decreases with the interdot space. This reduction of effective bandgap impacts the conversion efficiency of QDs SC.

Figure 2.13 shows that the energy level becomes less and more bounded minibands are located within the barrier with QDs radius from 2 to 5 nm

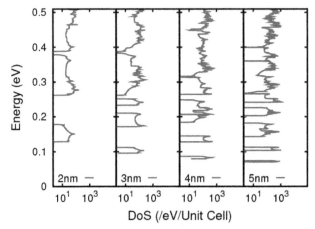

Figure 2.13 The DoSs for Ge/Si QDs square superlattice with QD thickness 4 nm, interdot space 0.3 nm and varied radius from 2 to 5 nm.

because of the weaker confinement in the *x–y* plane. With a larger radius, more minibands and excited states increase the probability of mixing of minibands since the excited states have better chances of wavefunction penetration and coupling between QDs. On the other hand, the bandwidth of minibands gets smaller with the QDs radius, which could lessen the probability of band mixing because the larger distance between QDs results in lesser interaction between QDs in the *x–y* plane. Thus, there are two competitive mechanisms affecting the effective bandgap of bulk material for the radius of QDs in superlattice.

In Figure 2.14, the energy levels become less with varied thickness from 2 to 8 nm because of the weaker confinement in *z*-direction as expected. Meanwhile, the bandwidth depends less on the thickness because the interaction between QDs is strongly related to the distance between QDs in the *x–y* plane and not to the thickness of QDs. The significant dependence of band structure on QDs symmetry is also revealed for the first excited miniband for a thickness of 4 nm, which is mixing of nondegenerate p-orbit along *z*-direction and two-fold degenerate p-orbit in the *x–y* plane, as shown in Figure 2.10. With the thickness changing from 4 to 8 nm, the nondegenerate p-orbit along *z*-direction hasmuch weaker confinement and separate to first excited band

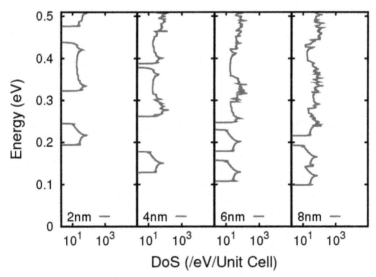

Figure 2.14 The DoSs for Ge/Si QDs square superlattice with QD radius 2 nm, interdot space 0.3 nm and varied thickness from 2 to 8 nm.

for a thickness of 6 nm and then mixes with the ground band for a thickness of 8 nm. On the other hand, because of the weaker confinement in z-direction, more minibands located within the barrier increase the probability of miniband mixing and reduce the effective bandgap of bulk material.

2.5.3 QDs IB SC

The IBSC consisted of Ge QDs with a radius of 2 nm and a thickness of 4 nm, and the varied interdot space embedded in the silicon matrix is considered. The blackbody radiation with 5778K (T_s in Equation (2.8)) is used to simulate sun illumination. The result is shown in Figure 2.15. In order to compare between single bandgap SC and QDs IBSC, using the same formula of Equations (2.7)–(2.10) without contribution of J_{IB} and $\mu_{CV} = V$ as single bandgap SC [23], we have bulk silicon exhibiting 53.5 mA/cm² short circuit current, 1.06 V open-circuit voltage, and 36.8% conversion efficiency under 1000 sun illumination ($\chi = 1000$).

The Ge QDs array embedded in bulk silicon contributes additional current by J_{IB} which increases with the density of QDs, as in Figure 2.15, because sub-bandgap photons are absorbed through two-photon transition from VB to CB to generate electron-hole pair. Thus, the conversion efficiency increases

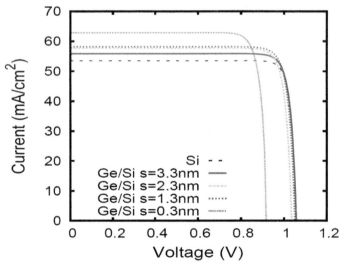

Figure 2.15   The simulated $J - V$ characteristics for Ge/Si QDs IBSC with interdot space from 3.3 to 0.3 nm under 1000 sun illumination. The simulated $J - V$ characteristics for silicon single bandgap solar cell is shown for comparison.

up to 38.9% with the density of QDs as high as $3.56 \times 10^{12}/cm^2$ (the interdot space 1.3 nm). However, the conversion efficiency turns around with QDs density $5.41 \times 10^{12}/cm^2$ as the interdot space decreases to 0.3 nm, as in Figure 2.16. It is a result of smaller open-circuit voltage in *J–V* characteristics. Figure 2.17 shows the effective bandgap voltage from the band structure in Figure 2.12 and open-circuit voltage from IBSC calculation in Figure 2.15 with the varied QDs density due to the interdot space. The reduced open-circuit voltage is explained by the reduction of effective bandgap revealed in Figure 2.12 that higher excited states become continuous states through miniband mixing due to stronger QDs interaction between each other as the interdot space decreases. This phenomenon is physically consistent that, as the density of Ge QDs is high, Ge QDs array approaches bulk Ge whose bandgap is 0.66 eV and open-circuit voltage is 0.63 V under detailed balance limit for a single bandgap SC calculated by Equations (2.7)–(2.10) without contribution of J_{IB} and $\mu_{CV} = V$.

In addition to the effect of interdot space on the conversion efficiency, the effect of radius and thickness is revealed in Figure 2.18. The conversion efficiency has a variation of more than 20% from 19.4 (radius 2 nm and thickness 8 nm) to 40.1% (radius 4 nm and thickness 2 nm) under interdot space 0.3 nm for the radius in range of 2–5 nm and the thickness in range of 2–8 nm. Meanwhile, the variation of conversion efficiency is

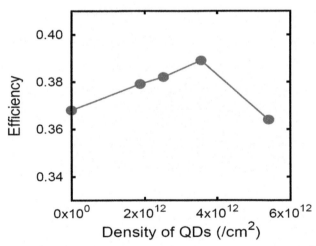

Figure 2.16 The conversion efficiency versus QDs density in the range of $10^{12}/cm^2$ due to varied interdot space from 3.3 to 0.3 nm for Ge/Si QDs IBSC under 1000 sun illumination. The data for zero QDs density is from silicon single bandgap solar cell.

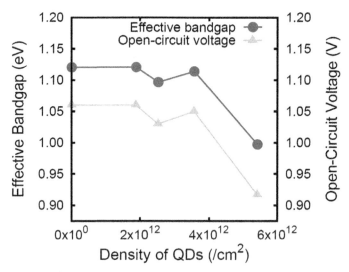

Figure 2.17   Plot of the effective bandgap voltage and open-circuit voltage versus QDs density in the range of $10^{12}/\text{cm}^2$ due to varied interdot space from 3.3 to 0.3 nm for Ge/Si QDs IBSC under 1000 sun illumination. The data for zero QDs density is from silicon single bandgap solar cell.

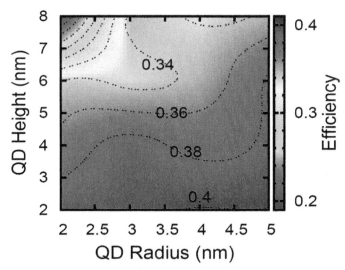

Figure 2.18   The conversion efficiency as a function of radius and thickness for the interdot space 0.3 nm of Ge/Si QDs IBSC under 1000 sun illumination. The maximum efficiency is around 40.1% for radius in the range of 4 to 5 nm and thickness 2 nm.

within 2% under the interdot space larger than 1.3 nm for the same range of radius and thickness (not shown). There are two factors responsible for this strong dependence of conversion efficiency on radius and thickness under the interdot space 0.3 nm. The first is the less confinement for thicker QDs. The last is the stronger QDs interaction for narrower interdot space and smaller radius. Less confinement and stronger interaction between QDs result in more possibility of miniband mixing and reduction of effective bandgap. Therefore, the conversion efficiency is down to 19.4% for QDs with the radius 2 nm and thickness 8 nm since the effective bandgap of bulk silicon is reduced to 0.83 eV. On the other hand, the QDs with the radius 4 nm and the thickness 2 nm are not impacted by QDs interaction between QDs due to larger distance but have conversion efficiency up to 40.1% due to higher volume ratio (larger radius) and QDs density (narrower space).

2.6 Conclusions

The FEM method by a unit cell with consideration of translation symmetry and appropriate boundary condition was developed to efficiently and accurately calculate the miniband structure and DoSs for an idealistic QDs array with realistic geometrical parameters. In a simple 1D array of square potential energy, this method matches well with the Kronig-Penney method. Without extra approximations, it can be simply extended to a 2D or 3D realistic QDs array. The example of an idealistic in-plane 3D Ge QDs array in bulk silicon matrix was used to calculate miniband structure and DoSs for guiding QD SC design by investigating the geometrical effect on the ultimate conversion efficiency of the IBSC. In future, the considerations of imperfect light trapping [24] by QDs geometrical structure, quantum efficiency, and nonradiative recombination [25] by QDs band structure without the ideal assumption in Section 2.4 are needed to provide a more realistic estimation on conversion efficiency for the QDs SC.

References

[1] A. Luque and A. Martì, "Increasing the efficiency of ideal solar cells by photon induced transitions at intermediate levels," Phys. Rev. Lett., vol. 78, pp. 5014–5017, 1997.

[2] A. Luque, A. Martì, and C. Stanley, "Understanding intermediate-band solar cells," Nature Photonics, vol. 6, pp. 146–152, 2012.

[3] A. S. Lin, W. Wang and J. D. Phillips "Model for intermediate band solar cells incorporating carrier transport and recombination," J. Appl. Phys., vol. 105, p. 064512, 2009.

[4] A. M. Kechiantz, L. M. Kocharyan, and H. M. Kechiyants, "Band alignment and conversion efficiency in Si/Ge type-II quantum dot intermediate band solar cells," Nanotechnology, vol. 18, p. 405401, 2007.

[5] Y. Okata, T. Morioka, K. Yoshida, R. Oshima, Y. Shoji, T. Inoue, and T. Kita, "Increase in photocurrent by optical transitions via intermediate quantum states in direct-doped InAs/GaNAs strain-compensated quantum dot solar cell," J. Appl. Phys., vol. 109, p. 024301, 2011.

[6] M. Igarashi, M. F. Budiman, W. Pan, W. Hu, N. Usami, and S. Samukawa, "Quantum dot solar cells using 2-dimensional array of 6.4-nm-diameter silicon nanodisks fabricated using bio-templates and neutral beam etching," Appl. Phys. Lett., vol. 101, p. 063121, 2012.

[7] Y. Tamura, T. Kaizu, T. Kiba, M. Igarashi, R. Tsukamoto, A. Higo, W. Hu, C. Thomas, M. E. Fauzi, T. Hoshii, I. Yamashita, Y. Okada, A. Murayama, and S. Samukawa, "Quantum size effects in GaAs nanodisks fabricated using a combination of the bio-templates technique and neutral beam etching," Nanotechnology, vol. 24, p. 285301, 2013.

[8] O. L. Lazarenkova and A. A. Balandin, "Miniband formation in a quantum dot crystal," J. Appl. Phys., vol. 89, no. 10, pp. 5509–5515, 2001.

[9] C.-W. Jiang and M. A. Green, "Silicon quantum dot superlattices: Modeling of energy bands, densities of states, and nobilities for silicon tandem solar cell applications," J. Appl. Phys., vol. 99, p. 114902, 2006.

[10] C.-H. Huang, X.-Y. Wang, M. Igarashi, A. Murayama, Y. Okada, I. Yamashita, and S. Samukawa, "Optical absorption characteristic of highly ordered and dense two-dimensional array of silicon nanodiscs," Nanotechnology, vol. 22, p. 105301, 2011.

[11] M. F. Budiman, W. Hu, M. Igarashi, R. Tsukamoto, T. Isoda, K. M. Itoh, I. Yamashita, A. Murayama, Y. Okada, and S. Samukawa, "Control of optical bandgap energy and optical absorption coefficient by geometric parameters in sub-10 nm silicon-nanodisc array structure," Nanotechnology, vol. 23, p. 065302, 2012.

[12] W. Hu, M. F. Budiman, M. Igarashi, M.-Y. Lee, Y. Li, and S. Samukawa, "Modeling miniband for realistic silicon nanocrystal array," Math. Comput. Model., vol. 58, pp. 306–311, 2013.

[13] T. Fujii, T. Okada, M. Syazwan, T. Isoda, H. Endo, M. Rahman, K. Ito, and S. Samukawa, "Germanium nano disk array fabrication by combination of bio template and neutral beam etching for solar cell application,"

in 2014 IEEE 40th Photovoltaic Specialist Conference (PVSC), 2014, pp. 1033–1036.

[14] W. Hu, M. M. Rahman, M.-Y. Lee, Y. Li, and S. Samukawa, "Simulation study of type-II Ge/Si quantum dot for solar cell applications," J. Appl. Phys., vol. 114, p. 124509, 2013.

[15] J.-H. Lee, T. Shishidou, and A. J. Freeman, "Improved triangle method for two-dimensional Brillouin zone integrations to determine physical properties," Phys. Rev. B, vol. 66, p. 233102, 2002.

[16] P. E. Blöchl, O. Jepsen, and O. K. Andersen, "Improved tetrahedron method for Brillouin-zone integrations," Phys. Rev. B, vol. 49, pp. 16223–16233, 1994.

[17] L. Cuadra, A. Martì, and A. Luque, "Influence of the overlap between the absorption coefficients on the efficiency of the intermediate band solar cell," IEEE Trans. Electron Devices, vol. 51, pp. 1002–1007, 2004.

[18] A. Luque, A. Martì, E. Antolìn, and C. Tablero, "Intermediate bands versus levels in non-radiative recombination," Physica B: Condensed Matter, vol. 382, pp. 320–327, 2006.

[19] K. Schick, E. Daub, S. Finkbeiner, and P. Würfel, "Verification of a generalized Planck law for luminescence radiation from silicon solar cells," Appl. Phys. A., vol. 54, pp. 109–114, 1992.

[20] R. de L. Kronig and W. G. Penney, "Quantum mechanics of electrons in crystal lattices," Proc. R. Soc. Lond. A, vol. 130, pp. 499–513, 1931.

[21] A. Richter, M. Hermle, and S. W. Glunz, "Reassessment of the limiting efficiency for crystalline silicon solar cells," IEEE J. Photovoltaics, vol. 3, no. 4, pp. 1184–1191, 2013.

[22] W. Hu, M. Igarashi, M.-Y. Lee, Y. Li, and S. Samukawa, "Realistic quantum design of silicon quantum dot intermediate band solar cells," Nanotechnology, vol. 24, p. 265401, 2013.

[23] W. Shockley and H. J. Queisser, "Detailed balance limit of efficiency of p–n junction solar cells," J. Appl. Phys., vol. 32, no. 3, pp. 510–1519, 1961.

[24] D. M. Callahan, J. N. Munday, and H. A. Atwater, "Solar cell light trapping beyond the ray optic limit," Nano Letters, vol. 12, no. 1, pp. 214–218, 2012.

[25] U. Rau, U. W. Paetzold, and T. Kirchartz, "Thermodynamics of light management in photovoltaic devices," Phys. Rev. B, vol. 90, p. 035211, 2014.

3

Sputtering Epitaxy of Si and Ge for Application to Solar Cells

Wenchang Yeh

Interdisciplinary Graduate School of Science and Engineering,
Shimane University, Matsue, Shimane 690-8504, Japan

Abstract

The sputtering method has important advantages in terms of larger-area applicability, low environment load, low cost, and safety. Therefore, it is suitable for application to solar cell (SC) fabrications. In this chapter, a novel Si/SiGe–Ge multijunction SC structure will be proposed, and sputtering epitaxy (SE) of Si and Ge on Si(100) substrate at under 360°C for application to SCs will be demonstrated. Film characteristics, which were evaluated by transmission electron microscope (TEM), Raman scattering, Hall measurement, and so on, were shown in relation to temperature of post-deposition forming gas annealing (FGA). The conductivity of SE films were controlled by the cosputtering of acceptor or donor impurities with Si or Ge and were controlled in the order of 10^{16} to 10^{20} cm^{-3} nip-Si thin-film SC on Si(100) substrate and wafer nip-Ge thin-film SC on Ge(100) substrate were fabricated and analyzed for the first time.

Keywords: SE, co-sputtering, SCs, Si, Ge.

3.1 Introduction

Silicon SCs have been the most commonly produced ones because of the abundant supply of raw materials, extensive infrastructure, and the range of technologies offered by the integrated-circuit and flat-panel-display industries. Figure 3.1a shows the AM1.5 photon flux spectrum as received on the Earth's surface [1], together with absorption coefficient α of Si and Ge [2, 3].

Green Photonics and Smart Photonics, 47–60.

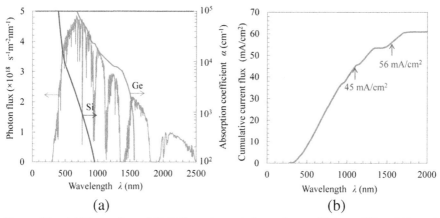

Figure 3.1   (a) Photon flux of AM1.5 spectra and absorption coefficient of Si and Ge and (b) Cumulative current flux in relation to absorption edge.

Assuming a 100% quantum efficiency of photon flux to carrier flux, maximum J_{max} was estimated in relation to wavelength of absorption edge, as shown in Figure 3.1b. α of Si dropped drastically at the absorption edge, resulting in a J_{max} of 45 mA/cm². This is one of the factors for the theoretical upper limit of Si SC at about 29.4% [4]. In order to increase conversion efficiency of Si-based SCs, it is effective to use Ge as a bottom cell; thus J_{max} can be further expanded to 56 mA/cm² because Ge has a direct band gap of 0.8 eV and an absorption edge of 1550 nm. However, there are two problems to be solved for realizing the Si/Ge multi-junction cell. First is the current mismatch between Si top cell and Ge bottom cell, and the second is the lattice mismatch between Si and Ge. For the first problem, photon flux between the top Si cell and bottom Ge cell should be equalized, however, is difficult when a thick Si wafer was used to fabricate the top Si cell. To solve this problem, we have proposed a novel Si/SiGe–Ge multijunction SC structure, as shown in Figure 3.2. In this structure, the bottom SiGe cell and the Ge cell were grown successively on the Si top cell, in which the SiGe cell and the Ge cell were series connected to rise open circuit voltage, and then the series-connected SiGe–Ge bottom cell was parallely connected with Si top cell. Here, PC matching between the Si top cell and the SiGe$_x$–Ge bottom cell is not necessary, but instead they are matched by open circuit voltage which is inherence for materials. The second problem is the lattice mismatch between Si and Ge. For actualizing this structure, the epitaxy of Si and Ge on Si substrate becomes important, and it will be discussed in the following section.

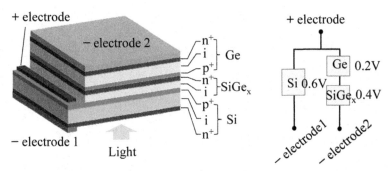

Figure 3.2 Proposed novel Si/SiGe-Ge multijunction SC structure.

3.2 SE of Si and Ge

Chemical vapor deposition [5–13] and MBE [14–17] was generally applied for the epitaxy of Si and Ge. However at temperatures of under 500°C the epitaxy rate in these methods was under 1 nm/s with a critical thickness of several hundreds of nms. Compared to CVD and MBE, SE [18–32] enables higher epitaxy rate with a larger critical thickness at low temperature because of enhancement of ad-atom migration caused by surface low kinetic energy particle bombardment such as ions [18] or electrons [33]. Furthermore, SE has other advantages of large area applicability, environment-friendly process gas, and low cost, so it holds out the hope for becoming a successful process for fabricating SCs. Bajor et al. [23] grew Ge film on Si(100) with a thickness of 1500 nm at 470°C by RF sputtering for the first time in 1982, in which the substrate potential was biased at −75 V compared to plasma potential to enhance ion bombardment, and Ohmi et al. [18] grew Si with a thickness of 400 nm at 350°C by RF sputtering, in which the substrate potential was biased at ∼10 V to optimize the ion bombardment. In recent years, SE without substrate bias was demonstrated successively; for example, in our study, 10 μm-thick Si and Ge was grown at >2 nm/s at <330°C by DC SE [25, 26, 31, 32, 42]. Here, the anode current I_a during Si epitaxy was measured in relation to anode bias voltage V_a. [26] As have shown in Figure 3.3, I_a was 184 mA at $V_a = 0$ V. I_a was kept constant as V_a was positively increased. Most of positive ions should be blocked to flow into anode in this condition, so I_a was composed of saturating electron current. As a result, electron flux was 4.1×10^{16} cm^{-2} s^{-1}. When V, was negatively biased to −18 V, I_a became 0, and further became negative value when V_a was further negatively increased, and then saturated at −0.934 mA at $V_a = -40$ V. Most of electrons should be blocked to flow into anode in this condition, so I_a was composed of saturating positive ion current.

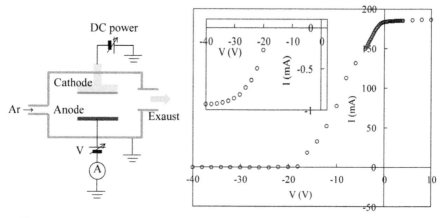

Figure 3.3   Anode current depending on anode bias voltage in sputtering system [26].

As a result, positive ion flux was 1.3×10^{14} cm^2 s^{-1}. The electron flux was two orders of magnitude larger than the ion flux, and, therefore, electrons should be the dominant species for surface bombardment in DC SE as $V_a = 0$ V. The Si atom flux at a deposition rate of 0.75 nm/s at 100 W was 3.7×10^{15} cm^{-2} s^{-1}, so a tenfold number of electrons were effectively bombarding the deposition surface. These may be the reason for the fast Si epitaxy at low temperature.

3.3 Characteristics of Si and Ge Films

Figure 3.4 shows the cross-sectional [110] HR TEM image of an interface between 2-μm thick epitaxial layer and Si(100) substrate. The lattice image of the area between the SE-Si film and the Si(100) wafer shows that the lattice image is matched without amorphous domain and stacking faults. Film quality was evaluated by Raman scattering spectroscopy (Raman) and RHEED in relation to substrate temperature T_{sub} and plasma power P. Raman revealed

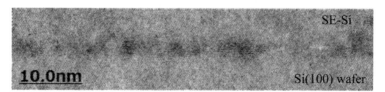

Figure 3.4   Cross-sectional [110] HR-TEM image of a interface between 2-μm thick epitaxial layer and Si(100) substrate.

that the film was amorphous at 107°C and became polycrystalline as the temperature was increased to 140°C and then became single crystalline at above 175°C. When P was at its standard condition, 100 W, the FWHM of Raman spectrum, W_R, was 5.20 cm^{-1}, but increased to 5.81 cm^{-1} with decreasing P to 25 W, and finally became completely amorphous at $P = 20$ W. TDD of SE-Si was evaluated by Sirtl etching and successive SEM observation [25]. TDD was $<10^4$ cm^{-2} at $P \geq 100$ W, but increased drastically to 1.0×10^6 cm^{-2} at $P = 75$ W, and became 8×10^6 cm^{-2} at $P = 25$ W. These results suggest that a higher P is essential for ensuring better crystal quality. From Hall measurement, the film showed n-type conduction although the target was non-doped, with electron mobility μ and concentration n of respectively 8×10^{16} cm^{-3} and 562 cm^2 V^{-1} s^{-1} for as-deposited film [31]. Concentration of oxygen and carbon impurities was measured by SIMS and was of the order of 10^{18} and 10^{17} cm^{-3}, respectively. The large n of 8×10^{16} cm^{-3} should be originated from interstitial oxygen which generates electrically active chains known as thermal donors [34, 35]. Since hydrogen neutralizes the oxygen-related thermal donors, n can be reduced by FGA. When FGA temperature, T_A, increased, the value of n started to decrease from $T_A = 600$°C and reached 1.2×10^{16} cm^{-3} at $T_A = 1000$°C [31]. At this condition, μ reached 1000 cm^2 V^{-1} s^{-1}, which is almost the same as that in bulk Si.

As for the SE-Ge on Si, Raman spectra of the SE-Ge films with thickness d of 100 and 2000 nm, on Ge (100) and on Si (100), were shown in Figure 3.5 together with that of a Ge wafer for comparison. A symmetric Raman shift at near $\sim$300.9 cm^{-1} corresponding to crystalline Ge, without the shoulder at lower wavenumbers, indicates that the films were single crystalline without amorphous and nanocrystalline phase [36, 37]. The Raman shift $\Delta\omega$ gives a degree of in-plane strain e [38]. Compared to $\Delta\omega$ of Ge wafer with 300.7 cm^{-1}, that of 100 nm-thick SE-Ge on Ge and on Si were respectively 300.7 and 301.7 cm^{-1}, and that of 2000 nm-thick SE-Ge on Ge and on Si were, respectively, 300.8 and 301.4 cm^{-1}. These results show that strain was not induced when SE-Ge film was formed on Ge, but compressive strain was induced when it was formed on Si(100). Besides, compressive strain relaxed gradually with increasing the thickness of SE-Ge film. Compared to the W_R of Ge wafer was 4.2 cm^{-1}, W_R of 100 nm-thick SE-Ge on Ge and on Si were, respectively, 4.8 and 5.5 cm^{-1}, and that of 2000 nm-thick SE-Ge on Ge and on Si were, respectively, 4.2 and 4.4 cm^{-1}. The crystal quality was better for SE-Ge films on Ge than that on Si when the film thickness was thin, but became similar to each other when the thickness was thick enough. Figure 3.6a, b show cross-sectional [110] TEM images of

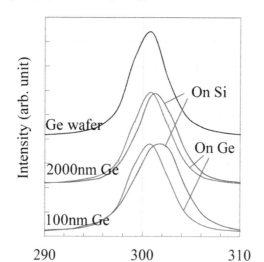

Figure 3.5   Raman spectra of Ge wafer, 100 nm SE-Ge on Ge(100), 100 nm SE-Ge on Si(100), 2000 nm SE-Ge on Ge(100), and 2000 nm SE-Ge on Si(100).

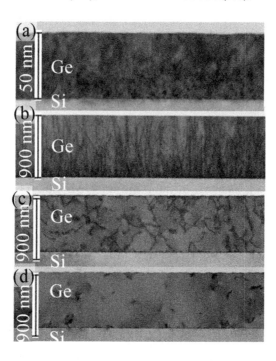

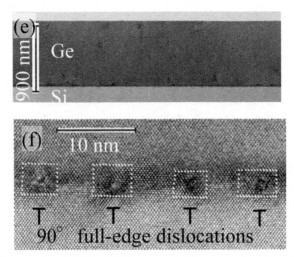

Figure 3.6 Cross-sectional [110] TEM images of, (a) as-deposited 50 nm-thick Ge film, (b) as-deposited 1000 nm-thick Ge film, (c)1000 nm-thick Ge film after 600°C annealing, (d) after 700°C annealing, and (e) after 800°C annealing. (f) Cross-sectional [110] HR-TEM image of (e) at the Ge–Si interface [42].

as-deposited SE-Ge films on Si with thickness of 50 and 1000 nm, respectively [42]. Surfaces were flat; however, dislocations and strain field exist within films. Figure 3.6(c–e) show cross-sectional TEM images of SE-Ge after FGA with T_A of 600, 700, and 800°C, respectively. The dislocations and strain field decreased considerably with increasing T_A; at 800°C, only misfit dislocations at the Ge–Si interface were observed, as shown in the cross-sectional [110] HR-TEM image in Figure 3.6f. Hall measurement was used to evaluate the electrical characteristics of Ge films. The conduction type of as-deposited SE-Ge film was p-type, although a non-doped Ge was used as sputtering target; therefore, holes were originated from defects. The Hall hole concentration p and mobility μ were 5×10^{16} cm^{-3} and 300 cm^2 V^{-1} s^{-1}, respectively. Figure 3.7 shows p and μ depending on T_A [42]. p and μ declined abruptly as T_A was increased from 400 to 440°C. The declining values should have resulted from the generation of dangling bonds because of desorption of gases in films. With further increase in T_A from 500 to 600°C, p and μ improved sharply; at $T_A = 700$°C, p and μ reached 1.3×10^{16} cm^{-3} and 1180 cm^2 V^{-1} s^{-1}, respectively. This improvement can be attributed to the removal of dislocations and decreased lattice-constant fluctuations. A further increase in T_A to 900°C resulted in roughening of the film surface, presumably owing to alloying between Ge and Si and a decrease in μ.

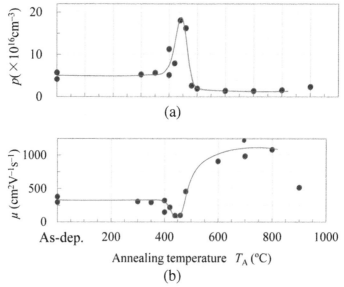

Figure 3.7 Annealing-temperature dependence of (a) hole concentration p and (b) mobility μ [42].

3.4 Conductivity Control of Epitaxial Ge and Si Films

It is easier to induce dopant into Si or Ge films in the SE method than in the MBE or the CVD methods, by means of co-sputtering dopant materials with Si or Ge. As for donor impurity, Sb is the best candidate since Sb is solid and less toxic. While for acceptor impurity, B, Al, and Ga are conceivable candidates.

For n-type SE-Si film, an Sb chip (99.999%) was placed on the Si target. The area of Sb on Si target was varied to change the donor concentration in the SE-Si film. As a result, n was controlled among 8×10^{16} cm^{-3} to 2.8×10^{20} cm^{-3}. The maximum n value was near as that of Sb-doped MBE-Si film grown at 300°C, 2.2×10^{20} cm^{-3} [39]. n was even larger than the solubility of Sb ($<2 \times 10^{19}$ cm^{-3}) in Si at <500°C [40], attributable to the non-equilibrium doping in this method. For p type SE-Si film, B, Al, or Ga were candidate for co-sputtering. However, the maximum hole concentration p didn't surpass 10^{20} cm^{-3} when Ga or Al were used for co-sputtering because of a lower solubility of these atoms in the Si. p was under the order of 10^{20} cm^{-3} when Ga or Al was used. When B was used for cosputtering, p reached a maximum of 2.6×10^{20} cm^{-3}.

For n-type SE-Ge film on Si(100), again Sb was co-sputtered with Si to induce donor impurity. The maximum n value was 8.4×10^{19} cm^{-3}. While for p-type SE-Ge film on Si(100), Al was co-sputtered with Ge. The maximum p was 1.0×10^{21} cm^{-3} [32].

Figure 3.8a, b, respectively, show the μ, n, p of i-Ge, p$^+$-Ge, and n$^+$-Ge at various film temperatures T [32, 42]. The hole mobility in i-Ge, μ_{pi}, increased with decreasing T as a function of $T^{-2.5}$ for $370 > T > 300$ K owing to acoustic and optical-phonon scattering [41]. Then, it decreased with decreasing T as a function of $T^{1.5}$ for $300 > T > 190$ K owing to ionized-defect scattering. The hole concentration in i-Ge, p_i, decreased as T decreased from 370 K $(1000/T = 2.7$ K$^{-1})$ to 190 K $(1000/T = 5.3$ K$^{-1})$. As T further decreased from 190 K $(1000/T = 5.3$ K$^{-1})$ to 100 K $(1000/T = 10$ K$^{-1})$, p_i increased again and saturated to 2×10^{16} cm^{-3} at $T < 100$ K. This can be explained by hopping conduction via trap state or carrier conduction in the acceptor band [42, 43]. For p$^+$-Ge and n$^+$-Ge films, the carrier concentration was, respectively, determined to be 9.8×10^{19} and 4.8×10^{19} cm^{-3} at 300 K. n_n and p_p were independent of T, indicating that these films were degenerated. [32]

3.5 SC Characteristics

Both Si and Ge thin film SC (TFSC) was fabricated by means of SE, as shown by the cross-sectional structure in Figure 3.9a, b, respectively. As for the Si TFSC, a 1000-nm thick intrinsic (i) SE-Si film and a 50-nm thick n$^+$

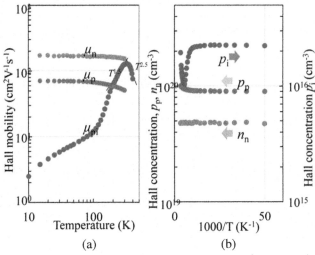

Figure 3.8   (a) Hall mobility and (b) Hall concentration of i-Ge, p$^+$-Ge, and n$^+$-Ge at various film temperatures (T) [32].

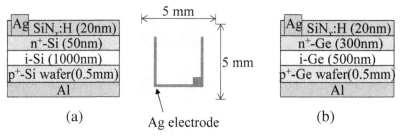

Figure 3.9   (a) Si TFSC structure and (b) Ge TFSC structure.

SE-Si film were grown on a low-resistivity p-Si(100) wafer with a resistivity of 0.002 Ωcm to form the n-type–intrinsic–p-type (nip) junction. The SE temperature was under 310°C. The substrate was heavily doped and had an extremely short lifetime, so the photogenerated carriers in the substrate did not contribute to the PC; only the carriers in the thin light-absorbing layer contributed to the PC. Therefore, this was in essence a TFSC, even though a Si wafer was used as a substrate. Then, the surface was capped with a 20 nm-thick SiN$_x$:H film as passivation film and was then annealed at 700°C for 30 min in a FG atmosphere at 1 atm. Next, Ag and Al pastes for the top and bottom electrodes were successively screen-printed on the top and bottom surfaces, respectively, and were then simultaneously sintered at 780°C for 30 s. The schematic front plan view of the surface Ag electrode was shown as an inset of Figure 3.9a. The surface Ag paste penetrated through SiN$_x$:H film and made contact with n$^+$ SE-Si film after sintering. The sample area was 5 × 5 mm^2. As for the Ge TFSC, except for the fact that the used substrate was a low-resistivity p-Ge(100) wafer with a resistivity of 0.22 Ωcm and the thickness of i layer and n$^+$ layer was, respectively, 500 and 300 nm which were grown at 360°C, the other structures and processes were the same as those in Si-TFSC. The relationship between the current density and the voltage (J–V) was measured under the illumination with an AM1.5G solar simulator, and the internal quantum efficiency IQE was measured (Oriel Instruments IQE200).

Figure 3.10a shows J–V characteristics of fabricated Si-TFSCs under AM1.5G sunlight illumination. [31] Figure 3.10a also shows the characteristics of a single-crystal Si bulk SC in which n$^+$ SE-Si was directly grown as an emitter on a Si wafer (as a base) [25]; the thickness and growth conditions of the n$^+$ Si emitter were completely identical to those for the TFSCs. As can be seen in the schematic cross-sectional structure of a bulk-Si SC in the inset of Figure 3.10a, the Si wafer served as the light-absorption layer. The short-circuit current density, J_{sc}; open-circuit voltage, V_{oc}; fill factor, FF; efficiency,

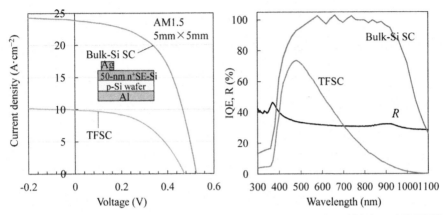

Figure 3.10 (a) *J–V* characteristics under AM1.5G and (b) *IQE* and *R* of fabricated Si-TFSC and Si bulk SC [31].

Table 3.1 Characteristics of Si TFSC and Si bulk-SC

	J_{sc} (mA/cm^2)	V_{oc} (V)	*FF* (%)	η (%)	R_{sh} (Ωcm^2)	R_s (Ωcm^2)
Si TFSC	9.9	0.47	50	2.30	650	14
Si Bulk-SC	23.8	0.53	55	6.9	1000	5

η; shunt resistance, R_{sh}; and series resistance, R_s, of both Si TFSC and Si bulk-SC were listed in Table 3.1. J_{sc} of TFSC was 9.6 mAcm^{-2}. Internal quantum efficiency, *IQE*, and reflectivity, *R*, of both Si TFSCs. and Si bulk SC were shown in Figure 3.10b. Here, the *R* of the two SCs was the same. We can focus on the PC generated at the SE-i layer, which was at a depth in the range of 50–1050 nm from the surface. Photons with an absorption depth (=1/α) below 1050 nm correspond to wavelengths λ below $\sim$500 nm. When the *IQE* of the TFSCs is compared with that of bulk-Si SCs, it can be seen, however, that *IQE* of the TFSCs at $\lambda < 500$ nm was still smaller than that of bulk-Si SCs. This suggests that the SE-i film quality was still lower than that of bulk Si and should be further improved to increase the performance of TFSCs. Although we have indicated that the μ of the SE-i film was almost the same as that of bulk Si, μ reflects the film performance of only majority carriers and not that of minority carriers. Further optimization of both annealing and passivation techniques is thus necessary to obtain much better characteristics of TFSCs. On the other hand, $J–V$ characteristics of Ge-TFSCs under AM1.5G sunlight illumination were shown in Figure 3.11(a). J_{sc}, V_{oc}, FF, η, R_{sh}, R_s, were listed in Table 3.2, and were 8.7 mA/cm^2, 153 mV, 0.42, 0.55%, 83 Ωcm^2, and 0.52 Ωcm^2, respectively. Figure 3.11b shows IQE and R as functions of

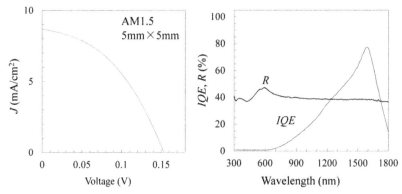

Figure 3.11 (a) *J–V* characteristics under AM1.5G and (b) *IQE* and *R*, of fabricated Ge-TFSC.

Table 3.2 Characteristics of GE TFSC

	J_{sc} (mA/cm^2)	V_{oc} (V)	*FF* (%)	η (%)	R_{sh} (Ωcm^2)	R_{s} (Ωcm^2)
Ge TFSC	8.7	0.15	42	0.55	83	0.52

wavelength. IQE was 76% at a maximum at 1600 nm, but decreased gradually with decreasing wavelength. IQE was almost 0 at wavelength lesser than 600 nm, suggesting active recombination centers at the near surface, and so passivation of these recombination centers is necessary for increasing J_{sc}.

3.6 Conclusion

A novel Si/SiGe–Ge multijunction SC structure was proposed, and the SE of Si and Ge on Si(100) substrate at under 360°C for application to SCs was demonstrated. Film characteristics, which were evaluated by TEM, Raman scattering, Hall measurement, and so on, were shown in relation to temperature of post-deposition forming gas annealing. The SE-Si and SE-Ge exhibited a good crystal quality and superior carrier conductance characteristics. For example, the maximum electron mobility achieved 1000 cm^2/Vs in SE-Si and maximum hole mobility achieved 1180 cm^2/Vs in SE-Ge. The conductivity of SE films was controlled by co-sputtering of acceptor or donor impurities with Si or Ge and was controlled in the order of 10^{16} to 10^{20} cm^{-3}. nip-Si thin-film SC on Si(100) substrate and wafer nip-Ge thin-film SC on Ge(100) substrate were fabricated and analyzed for the first time. The maximum IQE value exceeded 70% in both Si TFSC and Ge TFSC in which the SE method was used for forming the light absorption layer of SCs.

References

[1] http://rredc.nrel.gov/solar/spectra/am1.5/

[2] M. A. Green, Solar Energy Materials and Solar Cells, 92 (2008) 1305–1310.

[3] Pankove J. I. and P. Aigrain, Phys. Rev. 126, 3 (1962) 956–962.

[4] A. Richter, M. Hermle, S.W. Glunz, IEEE Journal of Photovoltaics 3 (4): 1184–1191. doi:10.1109/JPHOTOV.2013.2270351.

[5] A. Yamada, Y. Jia, M. Konagai, and K. Takahashi: J. Electron. Matter. 19 (1990) 1083.

[6] T. Hsu, B. Anthony, L. Breaux, R. Qian, S. Banerjee, and A. Tasch: J. Electron. Matter. 19 (1990) 1043.

[7] F. Mieno and Y. Furumura: J. Electron. Matter. 19 (1990) 1095.

[8] B. Anthony, T. Hsu, L. Breaux, R. Qian, and S. Banerjee: J. Electron. Matter. 19 (1990) 1089.

[9] A. Yamada, T. Oshima, M. Konagai and K. Takahashi: J. Electron. Matter. 24 (1995) 1511.

[10] J. Thiesen, E. Iwaniczko, K. M. Jones, A. Mahan, and R. Crandall: Appl. Phys. Lett. 72 (1999) 992.

[11] B. Cunningham, J. Chu, and S. Akbar, Appl. Phys. Lett. 59, 3574 (1991).

[12] M. Mäenpää, T. Kuech, M. Nicolet, S. Lau and D. Sadana, J. Appl. Phys. 53, 1076 (1982).

[13] Y. Tan and C. Tan, Thin Solid Films 520, 2711 (2012).

[14] B. Weir, B. Freer, R. Headrick, D. Eaglesham, G. Gilmer, J. Bevk, and L. Feldman: Appl. Phys. Lett. 59 (1991) 204.

[15] R. Malik, E. Gulari, S. H. Li, P. K. Bhattacharya, and J. Singh: J. Cryst. Growth 150 (1995) 984.

[16] Y. Ohmachi, T. Nishioka, and Y. Shinoda, J. Appl. Phys. 54, 5466 (1983).

[17] Y. Fukuda and Y. Kohama, Jpn. J. Appl. Phys. 26, L597 (1987).

[18] T. Ohmi, T. Ichikawa, T. Shibata, K. Matsudo, and H. Iwabuchi: Appl. Phys. Lett. 53 (1988) 45.

[19] G. Feng, M. Katiyar, N. Maley, and J. R. Abelson: Appl. Phys. Lett. 59 (1991) 330.

[20] T. Miyazaki and S. Adachi: J. Appl. Phys. 72 (1992) 5471.

[21] P. Sutter, C. Schwarz, E. Müller, V. Zelezny, S. Goncalves-Conto, and H. V. Känel: Appl. Phys. Lett. 65 (1994) 2220.

[22] J. Wang, H. Nakashima, J. Gao, K. Iwanaga, K. Furukawa, K. Muraoka, and Y. Sung: J. Vac. Sci. Technol. B 19 (2001) 333.

[23] G. Bajor, K. Cadien, M. Ray, J. Greene, and P. Vijayakumar, Appl. Phys. Lett. 40, 696 (1982).

[24] H. Hanafusa, Y. Suda, A. Kasamatsu, N. Hirose, T. Mimura, and T. Matsui, Jpn. J. Appl. Phys. 47, 3020 (2007).

[25] W. Yeh, K. Tatebe, K. Sugihara, and H. Huang, Jpn. J. Appl. Phys. 53, 025502 (2014).

[26] H. Huang and W. Yeh, Electrochem. Solid State Lett. 12 (3), H67 (2009).

[27] T. Ohmi, T. Ichikawa, T. Shibata, K. Matsudo, and H. Iwabuchi, Appl. Phys. Lett. 53, 45 (1988).

[28] M. Steglich, C. Patzig, L. Berthold, F. Schrempel, K. Füchsel, T. Höche, E. Kley and A. Tünnermann, AIP Advances 3, 072108 (2013).

[29] Z. Liu, X. Hao, A. Ho-Baillie, C. Tsao, M. A. Green, Thin Solid Films, 574, 99 (2015).

[30] H. Hanafusa, N. Hirose, A. Kasamatsu, T. Mimura, T. Matsui and Y. Suda, Jpn. J. Appl. Phys., 51, 055502 (2012).

[31] W. Yeh and K. Tatebe, Jpn. J. Appl. Phys. 54, 08KB07 (2015).

[32] W. Yeh, A. Matsumoto, and K. Sugihara, Jpn. J. Appl. Phys., 54, 08KD08 (2015).

[33] D. J. Stirland, Appl. Phys. Lett. 8, 326 (1966).

[34] H. Stein, S. Hahn, and S. Shatas, J. Appl. Phys. 59, 3495 (1986).

[35] N. Johnson and S. Hahn, Appl. Phys. Lett. 48, 709 (1986).

[36] Z. Iqbal, S. Veprek, A. Webb, and P. Capezzuto, Solid State Commun., 37, 993 (1981).

[37] T. Kanata, H. Murai, and K. Kubota, J. Appl. Phys., 61, 969 (1987).

[38] F. Cerdeira, C. Buchenauer, F. Pollak, and M. Cardona, Phys. Rev. B, 5, 580 (1972).

[39] M. Ohme, J. Werner, E. Kasper, J. of Crystal Growth, 310, p. 4531–4534. (2008).

[40] C. Claeys and E. Simoen, Germanium-Based Technologies-From Materials to Devices Elsevier, Amsterdam, 2007.

[41] D. Brown and R. Bray, Phys. Rev., 127, 1593 (1962).

[42] W. Yeh, A. Matsumoto, K. Sugihara and H. Hayase, ECS J. Solid State Sci. Technol. 3, (2014) Q195–Q199.

[43] C. S. Hung and J. Gliessman, Phys. Rev., 96, 1226 (1954).

4

Non-Stoichiometric SiC-based Solar Cells

Chih-Hsien Cheng, Yu-Chieh Chi and Gong-Ru Lin

Graduate Institute of Photonics and Optoelectronics, and
Department of Electrical Engineering, National Taiwan University,
Taipei, 10617, Taiwan, Republic of China

Abstract

All non-stoichiometric silicon carbide (Si_xC_{1-x})-based solar cell (SC) have
been demonstrated by using the plasma PECVD system. The C/Si composition
ratio of the non-stoichiometric Si_xC_{1-x} film grown at the RF plasma power
density of $110\,mW/cm^2$ is obtained as 0.503. The non-stoichiometric Si_xC_{1-x}
has a high-optical absorption coefficient of 2.1×10^5 cm^{-1} at the visible
region, which corresponds to the optical band gap energy of 2.04 eV. When
using the non-stoichiometric Si_xC_{1-x} with the C/Si composition ratio of
0.503 as the n-type and p-type layer, the non-stoichiometric Si_xC_{1-x}-based
p–n junction SC with the n-type Si_xC_{1-x} thickness of 50 nm obtains its
conversion efficiency of 0.37%. To enhance the conversion efficiency, the
non-stoichiometric Si_xC_{1-x}-based p–i–n junction SC by adding the intrinsic
Si_xC_{1-x} layer is demonstrated. Moreover, the series resistance of the non-
stoichiometric Si_xC_{1-x} film is optimized by adjusting the resistivities of
the n-type and p-type Si_xC_{1-x} film in order to enhance its conversion
efficiency. By doping with the PH$_3$ and B$_2$H$_6$ fluence ratios of 4.2 and 2.1%,
the resistivities of n-type and p-type non-stoichiometric Si_xC_{1-x} films are
improved to 0.87 and 0.12 Ωcm, respectively. On the other hand, the intrinsic
non-stoichiometric Si_xC_{1-x} thickness can further obtain the higher conversion
efficiency of devices. The non-stoichiometric Si_xC_{1-x}-based p–i–n junction
SC with the intrinsic non-stoichiometric Si_xC_{1-x} thickness of 25 nm has
the conversion efficiency of 1.7% with the filling factor (FF) of 0.25. When
combing the amorphous Si (a-Si)-based p–i–n junction SC to form the non-
stoichiometric Si_xC_{1-x}/a-Si-based tandem SC, its open-circuit voltage (V_{oc})

Green Photonics and Smart Photonics, 61–86.

and short-circuit PC density (J_{sc}) are acquired as 0.78 V and 19.1 mA/cm^2, respectively, which leads to the conversion efficiency and FF of 5.24% and 0.29. Finally, the intrinsic Si_xC_{1-x} film changes its C/Si composition by detuning the $[CH_4]/[CH_4+SiH_4]$ fluence ratio (R_{SiC}) to further enhance the conversion efficiency of the non-stoichiometric Si_xC_{1-x}-based SCs. By fixing the R_{SiC} at 0.3, the C/Si composition ratio of non-stoichiometric Si_xC_{1-x} films reduces to 0.36. The absorbance of non-stoichiometric Si_xC_{1-x} films grown with an R_{SiC} of 0.3 is enhanced to 3.8×10^5 cm^{-1}. Because of the increasing absorbance of the intrinsic Si_xC_{1-x} film, the V_{oc} and the J_{sc} of the non-stoichiometric Si_xC_{1-x}-based p–i–n junction SC are enhanced to 0.51 V and 19.7 mA/cm^2, respectively, which promote the conversion efficiency up to 2.24% with an FF of 0.283. For the non-stoichiometric Si_xC_{1-x}/a-Si tandem SC with the intrinsic Si_xC_{1-x} layer grown with the R_{SiC} of 0.3, it exhibits a highest conversion efficiency of 6.47% and an optimized FF of 0.332.

Keywords: Non-stoichiometric Si_xC_{1-x}, SC, p–n junction, p–i–n junction.

4.1 Introduction

The stoichiometric SiC has been comprehensively employed as the window layer for the group IV semiconductor-based optoelectronic devices over decades because the SiC material with chemical stability can easily be deposited at low temperature [1–5]. In general, the SiC film can be synthesized by versatile methods, such as HWCVD [6], Cat-CVD [7], magnetron cosputtering [8], MOCVD [9], and PECVD [10, 11]. For the solar cell (SC) application, the SiC exhibits high-optical band gap energy to assist the enhancement of the ultraviolet photodiode absorption, and such a unique feature also makes the conventional Si SCs a perfect visible blind window layer since early stage [12, 13]. In 2004, Myong et al. [14] reported that the amorphous silicon (a-Si)-based p–i–n junction SC by using a p-type a-SiC window layer can enhance its conversion efficiency to 11.2%. In 2007, Ogawa *et al.* [15] published a p–i–n junction SC with a conversion efficiency of 4.61% by using an n-type phosphorus-doped SiC window layer. Later on, Huang's group investigated that the microcrystalline silicon-based n–i–p junction SC by using an n-type microcrystalline SiC window layer promoted a conversion efficiency of up to 8.5% [16]. On the other hand, the relatively low-refractive index of SiC helps reduce the surface reflection of the Si-based SC, which improves the conversion efficiency of this device. Alternately, the thin-film SiC can be considered as a UV-absorptive and visible-blind absorption layer to combine with the other Si-based bi-layer or the tri-layer tandem SC, which effectively improves the solar power conversion at UV wavelengths without sacrificing the efficiency at visible wavelengths. However, the stoichiometric SiC has lower absorption coefficient and higher resistivity than Si [17, 18]; therefore, only few reports were emphasized for all stoichiometric SiC-based p–n or p–i–n junction SCs [19–23]. Gao *et al.* [19] reported that the a-Si$_{1-x}$C$_x$-based n–i–p junction SC employed as a semitransparent SC in an optical transmittance modulator only provides a conversion efficiency of less than 1%. Yunaz *et al.* [20] lately studied the hydrogenated intrinsic a-Si$_x$C$_{1-x}$ as the top cell of the triple junction SC. Lee *et al.* [21] changed the n-type Si$_x$C$_{1-x}$ thickness from 150 to 25 nm to decrease the series resistance of Si$_x$C$_{1-x}$-based p–n junction SCs, which also leads to a conversion efficiency enhanced from 5×10^{-3} to 4.7%. In addition, the SC with full-band absorption spectrum is applied to enhance the conversion efficiency; however, this device needs the absorbing layers with variable optical band gap to increase the absorbance [22, 23]. Therefore, the graded-index Si$_x$C$_{1-x}$-based multilayer tandem SC was developed to approach the higher conversion efficiency.

In order to overcome the low absorbance of the stoichiometric SiC, the non-stoichiometric Si$_x$C$_{1-x}$ materials have been successfully synthesized [24–26]. It exhibits larger absorption coefficient than the crystalline Si (c-Si) so as to perform a blind layer in visible light region. For the fabrication of the non-stoichiometric Si$_x$C$_{1-x}$, hydrogen was usually employed during the PECVD process because hydrogen passivation can improve the crystallinity of Si$_x$C$_{1-x}$ film [27]. However, the hydrogen passivation effect can contribute to the degraded absorption coefficient of Si$_x$C$_{1-x}$. The hydrogenated Si$_x$C$_{1-x}$ can be synthesized at low RF plasma condition under the hydrogen-free PECVD system because of the non-linear dissociation of silane (SiH$_4$) and methane (CH$_4$) gaseous mixture. Concurrently, it also enlarges its absorption coefficient, which is comparable to or even larger than crystalline SiC film [16]. Such an additional absorbance makes all non-stoichiometric Si$_x$C$_{1-x}$-based SCs enhance their conversion efficiency. Subsequently, several groups have applied the non-stoichiometric Si$_x$C$_{1-x}$ film as an absorbing layer to combine with a-Si-based SC to form the tandem SC. In 2014, Cheng *et al.* [28] varied the thickness of the intrinsic Si$_x$C$_{1-x}$ films from 100 to 25 nm for the Si$_x$C$_{1-x}$-based p–i–n junction SC. It also combined with a-Si-based p–i–n junction SC to provide the highest power conversion efficiency of up to 5.24%. Unfortunately, all non-stoichiometric Si$_x$C$_{1-x}$-based SCs and their characteristics were seldom discussed. In this chapter, the material characteristic of the non-stoichiometric Si_xC_{1-x} is studied. The discussion on physical mechanisms and designing considerations for such kinds of all non-stoichiometric Si_xC_{1-x}-based p–n and p–i–n junction SCs are emphasized. In addition, the combination of non-stoichiometric Si_xC_{1-x}-based SCs and a-Si-based SC to form the non-stoichiometric Si_xC_{1-x}/a-Si tandem solar is also observed.

4.2 All Non-Stoichiometric Si_xC_{1-x}-based p–n Junction SCs

In the first part of this chapter, all non-stoichiometric Si_xC_{1-x}-based p–n junction SCs are demonstrated. In contrast to the Si-based p–n junction SC, the non-stoichiometric Si_xC_{1-x}-based SC exhibits larger conversion efficiency because of its higher absorbance in visible light region. In addition, the absorption band edge of the Si_xC_{1-x} film can be extended to a long wavelength by detuning its C/Si composition ratio. The red-shifted cutoff wavelength promotes the solar energy conversion at the near-infrared region. On the other hand, the n-type Si_xC_{1-x} thickness can be changed to affect the series and the shunt resistances of the non-stoichiometric Si_xC_{1-x}-based p–n junction SC, which further enlarges the power conversion efficiency of the device.

4.2.1 The Synthesis of the Non-Stoichiometric Si_xC_{1-x}-based p–n Junction SC

The non-stoichiometric $Si_{1-x}C_x$ film was grown on the (100)-oriented Si and quartz substrates by using the hydrogen-free PECVD system with the reactance gas of SiH_4 and CH_4. The chamber pressure was set at 0.18 torr. The RF plasma power density was also controlled at 110 mW/cm^2. The fluence ratio defined as R_{SiC} = [CH$_4$]/[SiH$_4$]+[CH$_4$] was fixed at 0.5. The C/Si composition ratio of the non-stoichiometric Si_xC_{1-x} film and the relative composition ratio of chemical bonds of the non-stoichiometric Si_xC_{1-x} film were analyzed by using XPS with a Mg-Kα-line radiation source at 1253.6 eV. The optical transmittance and reflectance spectra were measured by using a Xeon light source and a monochromator. The n-type and p-type Si_xC_{1-x} films were grown by doping the reactant gas recipe with PH$_3$ and B$_2$H$_6$, respectively. Then, the postannealing treatment at 650°C was used to activate the dopants. The device structure of the non-stoichiometric Si_xC_{1-x}-based SC was defined as

ITO/p-Si$_x$C$_{1-x}$/n-Si$_x$C$_{1-x}$/Al. The thicknesses of the ITO anode and the Al cathode were 80 and 200 nm, respectively. The thickness of the p-type Si$_x$C$_{1-x}$ film was controlled at 50 nm with a dopant concentration of 1.2×10^{17} cm^{-3}. The n-type Si$_x$C$_{1-x}$ film exhibits a dopant concentration of 5×10^{16} cm^{-3}, and its thickness was also fixed as 50 nm. The current-voltage (I-V) response of the Si$_x$C$_{1-x}$-based SC was measured.

4.2.2 The XPS Analysis of Non-Stoichiometric Si$_x$C$_{1-x}$ Film

The C/Si composition ratio of non-stoichiometric Si$_x$C$_{1-x}$ grown at the RF plasma power density of 110 mW/cm^2 is measured by using the XPS analysis, as shown in Figure 4.1. By depositing at RF plasma power density of 110 mW/cm^2, the C content in non-stoichiometric Si$_x$C$_{1-x}$ film is obtained as 32.1% with a fraction index x of 0.67 for the Si$_x$C$_{1-x}$. This phenomenon is attributed to the larger energy for CH$_4$ decomposition because the CH$_4$ has a larger decomposed energy. Therefore, the C atoms hardly incorporate into Si$_x$C$_{1-x}$ under the RF plasma power density growth. Solomon *et al.* [29] revealed that the C content in Si$_x$C$_{1-x}$ film is enhanced by increasing the RF plasma power density to 300 (mW/cm^2). In more detail, the Si$_{(2p)}$

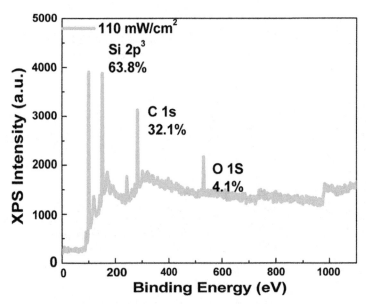

Figure 4.1 X-ray photoelectron spectrum of Si$_x$C$_{1-x}$ deposited by the RF plasma power density of 110 mW/cm^2.

orbital electron-related XPS signal for the non-stoichiometric Si_xC_{1-x} films grown at the RF plasma power density of 110 mW/cm^2 was analyzed by using three Gaussian-peak functions, as shown in Figure 4.2. The $Si_{(2p)}$ orbital electron-related XPS spectra are composed of three Gaussian peaks from Si–C-, C–Si–O-, and Si–Si-related bonds with their energies of 100.3, 101.75, and 99.5 eV, respectively [24]. At an RF plasma power density of 110 mW/cm^2, numerous C atoms are decomposed from CH$_4$ to combine with Si atoms, which forms Si–C bonds.

4.2.3 The Optical Absorption Analysis of the Non-Stoichiometric Si_xC_{1-x} Film

Figure 4.3 shows the absorption spectrum of the non-stoichiometric Si_xC_{1-x} films grown at 110 mW/cm^2 RF plasma power density, which exhibits the strongest optical absorption coefficient to 2.1×10^5 cm^{-1} at UV wavelength region. In comparison with c-Si, the non-stoichiometric Si_xC_{1-x} grown at the RF plasma power density of 110 mW/cm^2 has the larger optical absorption coefficient at the wavelength between 400 and 600 nm because of an additional absorbance from the buried Si–Si bonds in the non-stoichiometric Si_xC_{1-x} film. Moreover, the optical band gap of the non-stoichiometric Si_xC_{1-x} grown

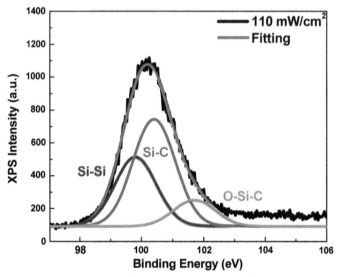

Figure 4.2　The $Si_{(2p)}$-related XPS spectra for the non-stoichiometric Si_xC_{1-x} films grown at the RF plasma power density of 110 mW/cm^2.

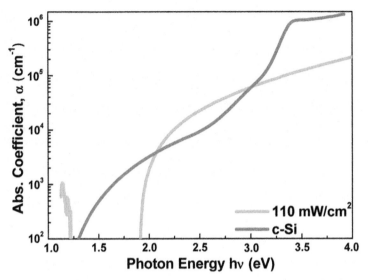

Figure 4.3   The optical absorption spectra of the c-Si and non-stoichiometric Si$_x$C$_{1-x}$ films grown at the RF plasma power density of 110 mW/cm^2.

at the RF plasma power density of 110 mW/cm^2 can be fitted by using Tauc's equation. From Tauc's equation, the optical band gap of the non-stoichiometric Si$_x$C$_{1-x}$ can be obtained as 2.04 eV. In comparison with the 3C–SiC (2.36 eV), the non-stoichiometric Si$_x$C$_{1-x}$ exhibits lower band gap because the excessive Si content in non-stoichiometric Si$_x$C$_{1-x}$ effectively decreases the optical band gap. Note that the concurrently mixed Si and SiC in non-stoichiometric Si$_x$C$_{1-x}$ film provides a higher optical absorption coefficient than the Si or stoichiometric SiC materials.

4.2.4 The PC Simulation of the Non-Stoichiometric Si$_x$C$_{1-x}$-based p–n Junction SC

Because the holes and electrons within penetration depth are separated by the build-in field, the PC density spectrum of the non-stoichiometric Si$_x$C$_{1-x}$ films can be evaluated from its absorption spectrum by employing the following equation:

$$I_{SC}(\lambda) = qA\gamma\alpha(\lambda)\Phi_0(\lambda)e^{-\alpha(\lambda)d}(L_n + L_p)$$
$$L_n = \sqrt{D_n\tau_n} = \sqrt{\frac{\mu_n KT}{qBN_a}},$$
$$L_p = \sqrt{D_p\tau_p} = \sqrt{\frac{\mu_p KT}{qBN_d}},$$
(4.1)

where q denotes the electron charge; A is the active area of the non-stoichiometric Si_xC_{1-x} SC; is the EQE, which is determined as 80% with the range between 450 and 900 nm; $\Phi_0(\lambda)$ is the flux density of the AM 1.5G light source; $\alpha(\lambda)$ is the absorption spectrum of the non-stoichiometric Si_xC_{1-x} films; d is the thickness of the non-stoichiometric Si_xC_{1-x} film; L_n and L_p are the diffusion lengths of the minority carriers in the neutral region, which are related to the electron and hole diffusion lengths in p-type and n-type Si_xC_{1-x}, respectively; D_n and D_p are the diffusion coefficients of minority carriers; τ_n and τ_p are the minority carrier lifetime in neutral region; K and T are the Boltzmann constant and temperature, respectively; N_a and N_d are the minority carrier concentration in p-type and n-type Si_xC_{1-x}, respectively; B is the radiative recombination coefficient, which is given as 1.5×10^{-12} cm^3/s; and μ_n and μ_p are the minority carrier concentrations in p-type and n-type Si_xC_{1-x}, respectively. Assuming that L_n and L_p are 83 μm and 93 μm, respectively, the PC density spectrum of the non-stoichiometric Si_xC_{1-x} grown at the RF plasma power of 110 mW/cm^2 under the AM 1.5G illumination is shown in Figure 4.4. After integrating by following Equation (4.2), the calculated short-circuit PC density (J_{sc}) of the non-stoichiometric Si_xC_{1-x} is acquired as 9.8 mA.

$$I_{SC} = \int_0^\infty I_{SC}(\lambda)d\lambda = qA\gamma(L_n + L_p)\int_0^\infty \alpha(\lambda)\Phi_0(\lambda)e^{-\alpha(\lambda)z}d\lambda, \quad (4.2)$$

4.2.5 The Performance of the Non-Stoichiometric Si_xC_{1-x}-based p–n Junction SC

In SC application, the AM 1.5G-illuminated I–V curve of the non-stoichiometric Si_xC_{1-x} p–n junction SC with 50-nm n-type Si_xC_{1-x} thicknesses is shown in Figure 4.5, which represents that the open-circuit voltage (V_{oc}) and the J_{sc} of the non-stoichiometric Si_xC_{1-x}-based p–n junction SC are 55 mV and 30.6 mA/cm^2, respectively. It provides a FF of 0.24 and a conversion efficiency of 0.37%. It can be explained by an equivalent circuit model of SCs with its PC (I_{pv}) described as

$$I_{pv} = I_L - I_s\left\{\exp\left[\frac{q(V + R_sI)}{nkT}\right] - 1\right\} - \frac{V + R_sI_{pv}}{R_{sh}}, \quad (4.3)$$

where I_L denotes the PC; I_s is the saturation current; n is the ideal factor of p–n junction diode; kT is the Boltzmann constant; V_{pv} is the voltage of the SC under illumination, respectively; and R_s and R_{sh} are the series and shunt resistance of

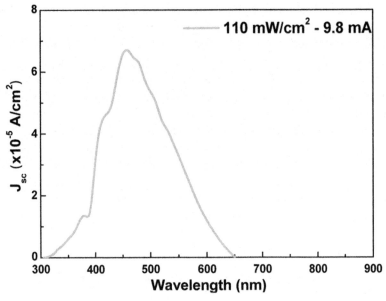

Figure 4.4 The PC density spectrum of non-stoichiometric Si$_x$C$_{1-x}$ films grown at the RF plasma power density of 110 mW/cm^2 under AM 1.5G illumination.

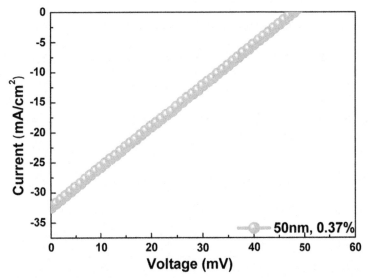

Figure 4.5 *I–V* characteristic of all non-stoichiometric Si$_x$C$_{1-x}$-based p–n junction SCs measured under AM-1.5 illumination.

the solar cell, respectively. The R_s and R_sh of the non-stoichiometric Si_xC_{1-x}-based p–n junction SC with an n-type Si_xC_{1-x} thickness of 50 nm can be simulated as 6 and 750 Ω, respectively, from the *I–V* response of the device.

According to the experiment result, the R_s of the non-stoichiometric Si_xC_{1-x} SC can affect its *I–V* characteristic, which further contributes to the variation of the conversion efficiency. By using Equation (4.3), the conversion efficiency of the non-stoichiometric Si_xC_{1-x}-based p–n junction SC as function of R_s can be simulated is shown in Figure 4.6. When the R_s decreases to 0.6 Ω, the conversion efficiency of the non-stoichiometric Si_xC_{1-x}-based p–n junction SC increases from 7×10^{-3} to 3%. Therefore, the R_s improvement can enhance the conversion efficiency. One of the methods for improving the R_s is optimizing the doping concentration of p-type and n-type Si_xC_{1-x} to decrease the resistivity. The related studies will be discussed in the next section.

4.3 All Non-Stoichiometric Si_xC_{1-x}-based p–i–n Junction SC

In the second part, the intrinsic Si_xC_{1-x} film is added into the non-stoichiometric Si_xC_{1-x}-based p–n junction SC to form the non-stoichiometric

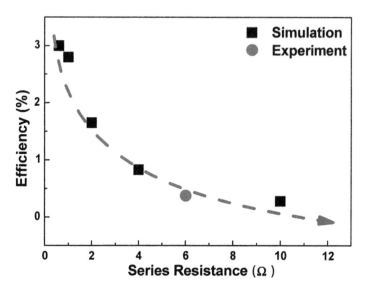

Figure 4.6   The simulated and experiment conversion efficiency of the non-stoichiometric Si_xC_{1-x}-based p–n junction SC as a function of R_s.

Si_xC_{1-x}-based p–i–n junction SC; thus, the absorbance of the device can be further enhanced. In addition, the intrinsic Si_xC_{1-x}-absorbing thickness is also optimized to increase the power conversion efficiency of the non-stoichiometric Si_xC_{1-x}-based p–i–n junction SC. The incomplete dissociation of the mixture gaseous occurred during growth helps reduce the defect density and improve the electrical properties of Si_xC_{1-x} films. Moreover, the electrical resistivities of n-type and p-type Si_xC_{1-x} films are optimized to further improve the power conversion efficiency by changing the PH_3 and the B_2H_6 dopant gas fluence ratios, respectively. After annealing, the V_{OC} and the J_{sc} of the non-stoichiometric Si_xC_{1-x} p–i–n junction SC are increased.

4.3.1 The Synthesis of the Non-Stoichiometric Si_xC_{1-x}-based p–i–n Junction SC

In this chapter, the n-type and p-type non-stoichiometric Si_xC_{1-x} films were fabricated by adding PH_3 and B_2H_6, respectively. By changing the doping fluence ratios defined as R_{B2H6} = $[B_2H_6]/[SiH_4+CH_4]$ and R_{PH3} = $[PH_3]/[SiH_4+CH_4]$ from 1.3 to 3.2% and from 2.5 to 8.3%, respectively, the Si_xC_{1-x}:P and Si_xC_{1-x}:B films were deposited so as to optimize their resistivities. Afterward, the PH_3 and B_2H_6 dopants were activated by post-annealing at 650°C for 30 min. The electrical properties of non-stoichiometric Si_xC_{1-x} films were measured by the four-probe measurements.

In application, Figure 4.7 shows the schematic diagram of the non-stoichiometric Si_xC_{1-x}/a-Si tandem SC. The p-type and the n-type Si_xC_{1-x} thicknesses were both controlled at 25 nm. The intrinsic Si_xC_{1-x} thickness fixed at 25 nm helps observe the conversion efficiency of non-stoichiometric Si_xC_{1-x}-based solar cells. A conversion efficiency of 3.97% and an FF of 0.66 for the bottom a-Si-based SC were observed.

4.3.2 Optimizing the Resistivity of n-Type and p-Type Non-Stoichiometric Si_xC_{1-x} Films by Detuning the PH_3- and B_2H_6-Doping Fluence Ratio in Reactant Gas Recipe

The n-type and p-type non-stoichiometric Si_xC_{1-x} films are detuned by changing the R_{PH3} and R_{B2H6}, respectively, to enhance the conversion efficiency of the SC. In Figure 4.8a, b, the resistivity of n-type Si_xC_{1-x}

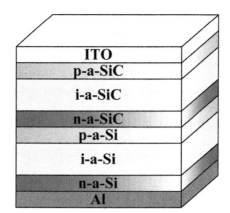

Figure 4.7 The device structure of the non-stoichiometric Si_xC_{1-x}/a-Si tandem solar cell.

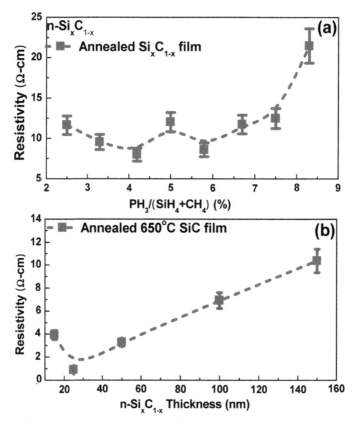

Figure 4.8 The resistivity of annealed n-type Si_xC_{1-x}:P films versus (a) doping fluence ratio and (b) thickness.

films reveals the response with parabolic curve when changing the R_{PH3} from 2.5% to 8.3%. After the 650°C annealing, the resistivity of n-type Si_xC_{1-x} films with its thickness of 100 nm is decreased to 8.5 Ω-cm when increasing the R_{PH3} to 4.2%. When further enlarging the R_{PH3}, the resistivity of the n-Si_xC_{1-x} is also increased. By fixing the R_{PH3} of 4.2%, the n-type Si_xC_{1-x} thickness is detuned to optimize to resistivity. The lowest resistivity of 0.875 Ω-cm for the n-Si_xC_{1-x} film is acquired when the n-Si_xC_{1-x} thickness is decreased to 25 nm.

In Figure 4.9a, the resistivity of p-type Si_xC_{1-x} films is deposited by detuning the R_{B2H6} from 1.3 to 3.2%. The resistivity of p-type Si_xC_{1-x} films is improved after annealing to 0.29 Ω-cm by increasing the R_{B2H6} to 2.1%.

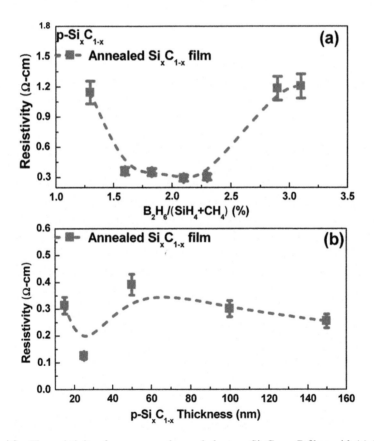

Figure 4.9   The resistivity of as-grown and annealed p-type Si_xC_{1-x}:B films with (a) doping density and (b) thickness.

When the p-type Si_xC_{1-x} thickness decreases to 25 nm, the lowest resistivity of the p-type Si_xC_{1-x} film is obtained as 0.12 Ω-cm in Figure 4.9b. The higher R_{PH3} and R_{B2H6} values contribute to the degraded resistivity of n-type and p-type Si_xC_{1-x} because the overdoped doping atoms in Si_xC_{1-x} films contribute to the carrier scattering and trapping effect [31, 32]. In addition, a similar phenomenon occurs when increasing the thickness. On the other hand, the higher thickness generates more defects in non-stoichiometric Si_xC_{1-x} films [33]. However, the resistivities of p-type and n-type Si_xC_{1-x} films are slightly degraded when the thickness to 15 nm due to the fewer dopant content in the thinner non-stoichiometric Si_xC_{1-x} film.

4.3.3 The Performance of the Non-Stoichiometric Si_xC_{1-x}-based p–i–n Junction SC

By optimizing the resistivities of n-type and p-type non-stoichiometric Si_xC_{1-x} films, the non-stoichiometric Si_xC_{1-x}-based p–i–n junction SC is performed under the AM-1.5G illumination. Because of the lower conversion efficiency of the non-stoichiometric Si_xC_{1-x}-based p–n junction SC, the intrinsic Si_xC_{1-x} layer is added because it can effectively broaden the absorption region to provide an additional absorbance [34]. Under AM-1.5G illumination, the $I–V$ curves of non-stoichiometric Si_xC_{1-x}-based p–i–n junction SCs without and with annealing are shown in Figure 4.10a, b. With the intrinsic Si_xC_{1-x} thickness of 25 nm, the V_{oc} and the J_{sc} of the as-grown non-stoichiometric Si_xC_{1-x}-based SC are measured as 0.45 V and 12.96 mA/cm^2, respectively. These provide the conversion efficiency of 1.36% with the FF around 0.25. After annealing to activate the dopant in the p-type and n-type Si_xC_{1-x} layer, the V_{oc} and the J_{sc} of the annealed non-stoichiometric Si_xC_{1-x}-based SC with the intrinsic Si_xC_{1-x} thickness of 25 nm are promoted

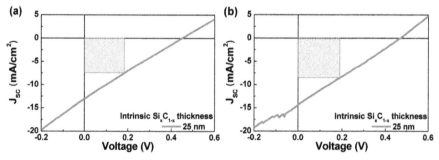

Figure 4.10   The illuminated $I–V$ curve of (a) as-grown and (b) annealed non-stoichiometric Si_xC_{1-x}-based p–i–n junction SCs with the intrinsic Si_xC_{1-x} thickness of 25 nm.

to 0.47 V and to 14.1 mA/cm^2, respectively. The conversion efficiency is further enhanced to 1.7% with the increasing FF around 0.27. Then, the non-stoichiometric Si_xC_{1-x}-based p–i–n junction combines with the a-Si-based p–i–n junction SC to form the tandem SC to acquire the higher conversion efficiency. Figure 4.11 shows the I–V curve of the non-stoichiometric Si_xC_{1-x}/a-SiC tandem solar under AM 1.5 G illumination. With the intrinsic Si_xC_{1-x} thickness fixed at 25 nm, the V_{oc} and the J_{sc} of the non-stoichiometric Si_xC_{1-x}/a-Si tandem SC are measured as 0.78 V and to 19.1 mA/cm^2, respectively, which contributes to a conversion efficiency of 5.24% with an FF of 0.29. Consequently, the higher conversion efficiency of the non-stoichiometric Si_xC_{1-x}-based p–i–n junction SC and the non-stoichiometric Si_xC_{1-x}/a-Si tandem SC are demonstrated because the intrinsic Si_xC_{1-x} layer provides an additional absorbance to generate the probability of the PC. In addition, the optimized p- and n-type Si_xC_{1-x} layers assist to promote the conversion efficiency. To obtain a higher conversion efficiency of the non-stoichiometric Si_xC_{1-x}-based SC, the depletion width or resistance is optimized by controlling the intrinsic Si_xC_{1-x} and the n-type and p-type Si_xC_{1-x} thickness. Moreover, the optical band gap of the non-stoichiometric Si_xC_{1-x} can be changed by controlling its C/Si compositional ratio in order to migrate the interfacial barrier. Fortunately, the C/Si composition ratio of the non-stoichiometric Si_xC_{1-x} can be finely detuned by precisely controlling the R_{SiC} during PECVD synthesis. These related studies will be discussed in the next section.

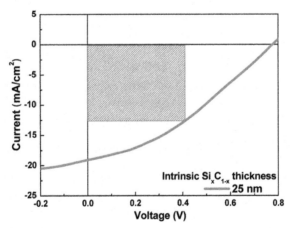

Figure 4.11 The illuminated I–V curve of non-stoichiometric Si_xC_{1-x}-based p–i–n junction SCs with intrinsic Si_xC_{1-x} thickness of 25 nm.

4.4 All Non-Stoichiometric Si_xC_{1-x}-based p–i–n Junction SCs with Lower C/Si Composition Ratio Grown Intrinsic Si_xC_{1-x} Layer

By changing the composition ratio of the intrinsic Si_xC_{1-x}-absorbing layer, the conversion efficiency of all the non-stoichiometric Si_xC_{1-x}-based p–i–n junction SC is further enhanced in the final part. Note that the absorbance of the Si_xC_{1-x} film in visible light region can be significantly enhanced by decreasing the C/Si composition ratio of the film, which is even one order of magnitude larger than that of the c-Si. Simultaneously, the V_{oc} and the J_{sc} of the non-stoichiometric Si_xC_{1-x}-based p–i–n junction SC are also enhanced to obtain a high FF. By reducing the fluence ratio and the hydrogen-free synthesis, the non-linear and incomplete dissociation of mixture gaseous is observed, which not only improves the surface passivation but also extends the band edge, enhances the absorption, and migrates the interfacial barrier of the nearly isolated and defect-free Si_xC_{1-x}. This further enhances the power conversion efficiency of the non-stoichiometric Si_xC_{1-x}-based p–i–n junction SC. By properly detuning the intrinsic Si_xC_{1-x} layer with optimized thickness and C/Si composition ratio, the non-stoichiometric Si_xC_{1-x}-based p–i–n junction SC exhibits a higher power conversion and FF than that of other structures.

4.4.1 The Fabrication of the Non-Stoichiometric Si_xC_{1-x}-based p–i–n Junction SC

The intrinsic non-stoichiometric Si_xC_{1-x} film with lower C/Si composition ratio was deposited by using an R_{SiC} of 0.3 during PECVD growth. The configuration of the non-stoichiometric Si_xC_{1-x}-based p–i–n junction SC and the non-stoichiometric Si_xC_{1-x}/a-Si tandem SC were the same as those of the device mentioned in previous section, as shown in Figure 4.7. The 25-nm thick p-type and n-type non-stoichiometric Si_xC_{1-x} films with resistivities of 0.12 and 0.87 Ω-cm, respectively, were chosen by doping the fixed R_{B2H6} and R_{PH3} to obtain a higher conversion efficiency of devices.

4.4.2 The XPS Analysis on the Composition of the Non-Stoichiometric Si_xC_{1-x} Films

Figure 4.12 shows the XPS spectrum of the non-stoichiometric Si_xC_{1-x} films grown with an R_{SiC} of 0.3. The non-stoichiometric Si_xC_{1-x} film reveals its C/Si composition ratio of 0.36 with a corresponding fraction index x of 0.74. Moreover, the O/Si composition ratio further decreases to maintain the quality of the non-stoichiometric Si_xC_{1-x} film. Figure 4.13 exhibits the $Si_{(2p)}$

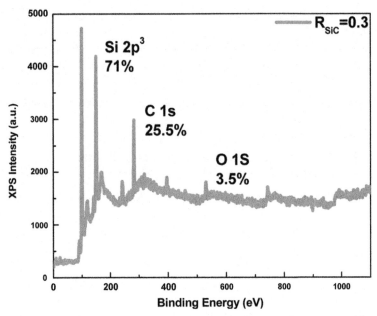

Figure 4.12 The XPS spectrum of non-stoichiometric Si$_x$C$_{1-x}$ films grown with an R_{SiC} of 0.3.

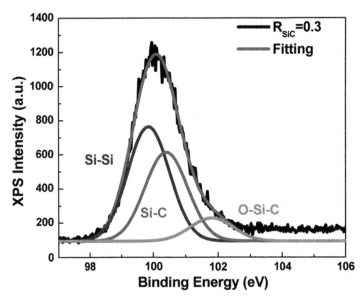

Figure 4.13 The Si$_{(2p)}$-related XPS spectrum for the non-stoichiometric Si$_x$C$_{1-x}$ films grown with an R_{SiC} of 0.3.

orbital electron-related XPS spectra, which can characterize the compositional bonds in non-stoichiometric Si_xC_{1-x} films. The $Si_{(2p)}$ orbital electron-related XPS spectrum is also deconvoluted by three separated Gaussian components originated from the contribution of the Si–Si, Si–C, and C–Si–O bonds, respectively. For the non-stoichiometric Si_xC_{1-x} film grown with an R_{SiC} of 0.3, the Si–Si-related bonds increase, which contributes to a red and near-infrared absorbance.

4.4.3 The Optical Absorption Analysis and PC Simulation of the Non-Stoichiometric Si_xC_{1-x} Films

The absorption spectrum of the non-stoichiometric Si_xC_{1-x} films grown with an R_{SiC} of 0.3 is shown in Figure 4.14. The non-stoichiometric Si_xC_{1-x} film grown with an R_{SiC} of 0.3 has the most broadened absorption spectrum with the highest absorption coefficient of up to 3.8×10^5 cm^{-1} at UV wavelength region. Moreover, the increasing Si–Si-related bonds in the non-stoichiometric Si_xC_{1-x} film can provide the additional absorbance at red and near-infrared regions, which induces the absorption band edge to 800 nm. This phenomenon contributes to the non-stoichiometric Si_xC_{1-x} film obtaining a low-optical band gap of 1.45 eV. Figure 4.15 shows the simulated PC density spectrum of the non-stoichiometric Si_xC_{1-x} under the AM1.5G illumination. With the non-stoichiometric Si_xC_{1-x} grown with an R_{SiC} of 0.3, the total J_{sc} of the

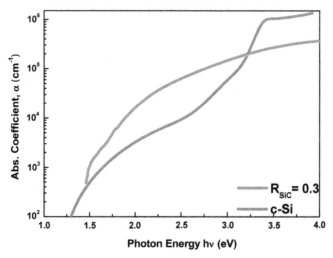

Figure 4.14 The absorption spectra of c-Si and non-stoichiometric Si_xC_{1-x} films grown with an R_{SiC} of 0.3.

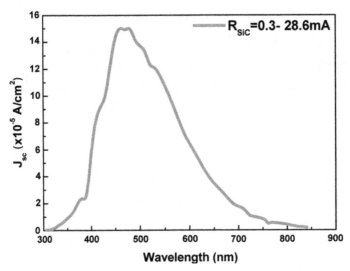

Figure 4.15 The PC spectrum of non-stoichiometric Si$_x$C$_{1-x}$ films grown with an R_{SiC} of 0.3 under AM 1.5G illumination.

non-stoichiometric Si$_x$C$_{1-x}$ is evaluated as 28.6 mA/cm^2 due to the strong absorbance at red and near-infrared regions.

4.4.4 The Performance of Non-Stoichiometric Si$_x$C$_{1-x}$-based SCs

The non-stoichiometric Si$_x$C$_{1-x}$-based p–i–n junction SC under the AM 1.5G illumination is characterized. In this case, the intrinsic Si$_x$C$_{1-x}$-absorbing layer with lower C/Si composition ratio increases an additional absorbance to generate the excited PC. With an intrinsic Si$_x$C$_{1-x}$-absorbing layer grown with an R_{SiC} of 0.3, the V_{oc} and the J_{sc} of the non-stoichiometric Si$_x$C$_{1-x}$-based p–i–n junction SCs are acquired as 0.51 V and 19.7 mA/cm^2, respectively, as shown in Figure 4.16a. It leads to a conversion efficiency of 2.24% with a corresponding FF of 0.283 because the intrinsic Si$_x$C$_{1-x}$-absorbing layer exhibits low series resistance. Finally, the non-stoichiometric Si$_x$C$_{1-x}$/a-Si tandem SC is synthesized to obtain the highest conversion efficiency. The FF and the conversion efficiency of the non-stoichiometric Si$_x$C$_{1-x}$/a-Si tandem SC can be further acquired as 0.332 and 6.47%, respectively, as shown in Figure 4.16b. The non-stoichiometric Si$_x$C$_{1-x}$ film with lower C/Si composition ratio has stronger absorbance and smaller R_s to enhance the overall performance of the non-stoichiometric Si$_x$C$_{1-x}$/a-Si tandem SC.

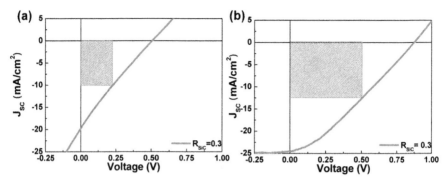

Figure 4.16　The AM1.5G illuminated I–V curves of the (a) non-stoichiometric Si_xC_{1-x}-based p–i–n junction and (b) tandem non-stoichiometric Si_xC_{1-x}/a-Si SCs with its i-Si_xC_{1-x}-absorbing layer grown with an R_{SiC} of 0.3.

Table 4.1 shows the list of Si_xC_{1-x}-based SCs and Si_xC_{1-x}/Si hybrid SCs with its performance specifications from different groups in recent years. For the single SC application, the structure of these devices almost resembles the Si_xC_{1-x}/Si hybrid structure. In general, the Si_xC_{1-x} combines with a-Si to form the SC with its conversion efficiency of 4–6%. When the a-Si replaces to c-Si, the conversion efficiency of devices can be further improved to 14–19% because the c-Si has better mobility and higher absorbance at

Table 4.1　Si_xC_{1-x}-based SCs and Si_xC_{1-x}/Si hybrid SCs in recent years

Materials	Device Structure	Performance Specifications	Reference
Si-nanocrystal: SiC/c-Si	Single SC Al grid/p-Si-nanocrystal:SiC/ n-c-Si/Ti/Al	V_{OC}: 0.463 V I_{sc}: 19.0 mA/cm^2 FF: 0.53 Eff: 4.66%	Qu *et al.* [4]
μc-3c-SiC/a-Si/ μc-Si_xC_{1-x}	Single SC TCO/TiO$_2$/n-μc-3C– SiC:H/i-a-Si:H/p-μc- SiC$_x$/Al	V_{OC}: 0.947 V J$_{sc}$: 8.31 mA/cm^2 FF: 0.59 Eff: 4.61%	Ogawa *et al.* [15]
a-SiC/a-Si	Single SC SnO$_2$:F/p-a-SiC:H/i-a- SiC:H/n-a- Si:H/ZnO:B/Ag	V_{OC}: 0.99 V J_{sc}: 8.56 mA/cm^2 FF: 0.71 Eff.: 6.04%	Yunaz *et al.* [20]
a-SiC/a-Si	Single SC TCO/p-a-SiC/ i-a-SiC/n-a-Si/Al	V_{OC}: 0.71 V J_{sc}: 11 mA/cm^2 FF: 0.56 Eff: 4.41%	Abdelkrim *et al.* [35]

Table 4.1 Continued

Materials	Device Structure	Performance Specifications	Reference
a-Si/-n-Si/μc-3C-SiC	Single SC Al/p-a-Si:H/p-c-Si/ n-μc-3C–SiC:H/ITO/Al	V_{OC}: 0.56 V J_{sc}: 35 mA/cm^2 FF: 0.724 Eff: 14:20%	Banerjee *et al.* [36]
nc-3C-SiC/c-Si/Si$_{1-x}$O$_x$	Single SC ITO/n-nc-3C-SiC: H/p-c-Si/p-μc-Si$_{1-x}$O$_x$:H/Al	V_{OC}: 0.68 V I_{sc}: 36.6 mA/cm^2 FF: 0.769 Eff: 19.1%	Irikawa *et al.* [37]
a-SiC	Single SC ITO/p-a-SiC/ n-a-SiC/Al	V_{OC}: 0.4 V I_{sc}: 32.7 mA/cm^2 FF: 0.32 Eff: 4.7%	Lee *et al.* [21]
a-SiC/a-Si	Tandem SC ITO/p-a-SiC/i-a-SiC/ n-a-SiC/p-a-Si:H/ i-a-Si:H/n-a-Si:H/Al	V_{OC}: 0.78 V I_{sc}: 19.1 mA/cm^2 FF: 0.35 Eff: 5.24%	Cheng *et al.* [28]
a-SiC/a-Si	Tandem SC SnO$_2$:F/p-a-SiC:H/ i-a-SiC:H buffer/i-a-SiC:H/ n-a-Si:H/ p-a-SiC:H/p-a-SiC:H buffer/i-a-Si:H/ n-a-Si:H/ZnO/Ag/Al	V_{OC}: 1.72 V I_{sc}: 5.18 mA/cm^2 FF: 0.67 Eff: 6.00%	Yunaz *et al.* [38]
a-Si$_x$C$_{1-x}$ a-Si$_x$C$_{1-x}$/a-Si	Single SC ITO/p-a-SiC/i-a-SiC/ n-a-SiC/Al	Single SC V_{OC}: 0.51 V J_{sc}: 19.7 mA/cm^2 FF: 0.284 Eff: 2.24%	[This work]
	Tandem SC ITO/p-a-SiC/i-a-SiC/ n-a-SiC/p-a-Si:H/ i-a-Si:H/n-a-Si:H/Al	Tandem SC V_{OC}: 0.82 V J_{sc}: 19.7 mA/cm^2 FF: 0.332 Eff: 6.47%	

near-infrared region. In comparison with these structures, the conversion efficiency of the non-stoichiometric Si$_x$C$_{1-x}$-based SC is only half lower than that of the Si$_x$C$_{1-x}$/Si hybrid SCs. However, for the tandem solar application, the non-stoichiometric Si$_x$C$_{1-x}$-based SC as the top cell can enhance its conversion efficiency. Compared to the previous works, the non-stoichiometric Si_xC_{1-x} is employed as not only the window layer but also the absorption layer to further enhance the conversion efficiency of SCs.

4.5 Conclusion

All Si_xC_{1-x}-based SCs are successfully demonstrated by synthesizing the non-stoichiometric Si_xC_{1-x} film. The non-stoichiometric Si_xC_{1-x} can improve its absorption coefficient and broaden its absorption spectrum. With the non-stoichiometric Si_xC_{1-x} grown at an RF plasma power density of 110 mW/cm^2, the optical band gap of 2.05 eV for the non-stoichiometric Si_xC_{1-x} film is obtained. When the n-type Si_xC_{1-x} thickness is fixed at 50 nm, the non-stoichiometric Si_xC_{1-x}-based p–n junction SC has a J_{sc} of 30.6 mA/cm^2 and a V_{OC} of 55 mV, resulting in a conversion efficiency of 0.37% with an FF of 0.24. To enhance the conversion efficiency, the intrinsic Si_xC_{1-x} layer is added into the SC to increase the absorbance. In addition, optimizing the resistivity of the p-type and n-type Si_xC_{1-x} films further raises the conversion efficiency. After thermal annealing, the optimized resistivities of 25-nm thick n-type and p-type Si_xC_{1-x} films are acquired as 0.875 and 0.12 Ω-cm by doping with the R_{PH3} and R_{B2H6} of 4.2 and 2.1%, respectively. When adding the 25-nm thick intrinsic Si_xC_{1-x}, both the non-stoichiometric Si_xC_{1-x}-based p–i–n junction and the tandem SCs increase their conversion efficiencies. The V_{OC} and the J_{sc} of the non-stoichiometric Si_xC_{1-x}/a-Si-based tandem SC with an intrinsic Si_xC_{1-x} thickness of 25 nm are measured as 0.78 V and 19.1 mA/cm^2, respectively, providing a conversion efficiency of 5.24% with a corresponding FF of 0.29. Finally, the intrinsic Si_xC_{1-x} film changes its C/Si composition by varying the R_{SiC} to enhance the absorbance of the non-stoichiometric Si_xC_{1-x}-based SCs. Because of the numerous Si content in non-stoichiometric Si_xC_{1-x}, the C/Si composition ratio of non-stoichiometric Si_xC_{1-x} films grown with an R_{SiC} of 0.3 also acquires as 0.36, which contributes to the lower optical band gap of 1.45 eV. With the R_{SiC} fixed as 0.3, the V_{OC} and the J_{sc} of non-stoichiometric Si_xC_{1-x}-based p–i–n junction SCs are measured as 0.51 V and 19.7 mA/cm^2, respectively, also prompting its conversion efficiency of 2.24% with an FF of 0.283. For the non-stoichiometric Si_xC_{1-x}/a-Si tandem SC with the intrinsic Si_xC_{1-x} layer grown with an R_{SiC} of 0.3, the FF and conversion efficiency can be up to 0.332 and 6.47%, respectively. Therefore, the non-stoichiometric Si_xC_{1-x} is employed not only as the window layer but also as the absorption layer to further enhance the conversion efficiency of SCs.

References

[1] Cheng CH, Wu CL, Chen CC, Tsai LH, Lin YH, Lin GR. Si-rich $Si_x C_{1-x}$ light-emitting diodes with buried Si quantum dots. IEEE Photonics J. 2012; **4**:1762–1775. DOI: 10.1109/JPHOT.2012.2215917.

[2] Chang PK, Hsieh PT, Lu CH, Yeh CH, Houng MP. Development of high efficiency p-i-n amorphous silicon solar cells with the p-μc-Si: H/p-a-SiC:H dual window layer. Sol. Energy Mater. Sol. Cells. 2011; **5**: 2659–2663. DOI: 10.1016/j.solmat.2011.05.036.

[3] Song BS, Yamada S, Assno T, Noda S. Demonstration of two-dimensional photonic crystals based on silicon carbide. Opt. Express. 2011; **19**: 11084–11089. DOI: 10.1364/OE.19.011084.

[4] Qu Y, Jokubavicius V, Hens P, Kaiser M, Wellmann P, Yakimova R, Syväjärvi M, Qu H. Broadband and omnidirectional light harvesting enhancement of fluorescent SiC. Opt. Express. 2012; 20: 7575–7579. DOI: 10.1364/OE.20.007575.

[5] Song D, Cho EC, Conibeer G, Flynn C, Huang Y, Green MA. Structural, electrical and photovoltaic characterization of Si nanocrystals embedded SiC matrix and Si nanocrystals/c-Si heterojunction devices. Sol. Energy Mater. Sol. Cells. 2008; **92**: 474–481. DOI: 10.1016/j.solmat.2007.11.002.

[6] Swain BP, Dusane RO. Effect of filament temperature on HWCVD deposited a-SiC-H. Mater. Lett. 2006; **60**: 2915–2919. DOI: 10.1016/ j.matlet.2005.10.050.

[7] Shimkunas AR, Mauger PE, Bourget LP, Post RS, Smith L, Davis RF, Wells GM, Cerrina F, Mcintosh RB. Advanced electron cyclotron resonance chemical vapor deposition SiC coatings and x-ray mask membranes. J. Vac. Sci. Technol. B. 1991; **9**: 3258–3261. DOI: 10.1116/ 1.585299.

[8] Kerdiles S, Rizk R, Pérez-Rodríguez A, Garrido B, González-Varona O, Calvo-Barrio L, Morante JR. Magnetron sputtering synthesis of silicon-carbon films: structural and optical characterization. Solid-State Electron. 1998; **42**: 2315–2320. DOI: 10.1016/S0038-1101(98)00232-9.

[9] Lim DC, Jee HG, Kim JW, Moon JS, Lee SB, Choi SS, Boo JH. Deposition of epitaxial silicon carbide films using high vacuum MOCVD method for MEMS applications. Thin Solid Films. 2004; **459**: 7–12. DOI: 10.1016/j.tsf.2003.12.140.

[10] Prado RJ, DAddio TF, Fantini MCA, Pereyra I, Flank AM. Annealing effects of highly homogeneous a-$Si_{1-x}C_x$:H. J. Non-Cryst. Solids. 2003; **330**: 196–215. DOI: 10.1016/S0022-3093(03)00526-X.

[11] Kurokawa Y, Tomita S, Miyajima S, Yamada A, Konagai M. Photolumi-nescence from silicon quantum dots in Si quantum dots/amorphous SiC superlattice. Jpn. J. Appl. Phys. 2007; **46**: L833–L835. DOI: 10.1143/JJAP.46.L833.

[12] Tawada Y, Kondo M, Okamoto H, Hamakawa Y. Hydrogenated amor-phous silicon carbide as a window material for high efficiency a-Si solar cells. Sol. Energy Mater. 1982; **6**, 299–315. DOI: 10.1016/0165-1633(82)90036-3.

[13] Baik SJ, Kang SJ, Lim KS. Extraction of carriers photogenerated at p type amorphous SiC window layer in amorphous Si solar cells. Appl. Phys. Lett. 2010; **97**: 122102. DOI: 10.1063/1.3491164.

[14] Myong SY, Kim SS, Lim KS. Improvement of pin-type amorphous sili-con solar cell performance by employing double silicon-carbide p-layer structure. J. Appl. Phys. 2004; **95**: 1525–1531. DOI: 10.1063/1.1639140.

[15] Ogawa S, Yoshida N, Itoh T, Nonomura S. Heterojunction amorphous silicon solar cells with n-type microcrystalline cubic silicon carbide as a window layer. Jpn. J. Appl. Phys. 2007; **46**: 518–522. DOI: 10.1143/JJAP.46.518.

[16] Huang Y, Dasgupta A, Gordijn A, Finger F, Carius R. Highly transparent microcrystalline silicon carbide grown with hot wire chemical vapor deposition as window layers in n-i-p microcrystalline silicon solar cells. Appl. Phys. Lett. 2007; **90**: 203502. DOI: 10.1063/1.2739335.

[17] Orpella A, Vetter M, Ferré R, Martín I, Puigdollers J, Voa C, Alcubilla R. Phosphorus-diffused silicon solar cell emitters with plasma enhanced chemical vapour deposited silicon carbide. Sol. Energy Mater. Sol. Cells. 2005; **87**: 667–674. DOI: 10.1016/j.solmat.2004.08.021.

[18] Demichelis F, Pirri CG, Tresso E. Influence of doping on the structural and optoelectronic properties of amorphous and microcrystalline silicon carbide. J. Appl. Phys. 1992; **72**: 1327–1333. DOI: 10.1063/1.351742.

[19] Gao W, Lee SH, Bullock J, Xu Y, Benson DK, Morrison S, Branz HM. First a-SiC:H photovoltaic-powered monolithic tandem electrochromic smart window device. Sol. Energy Mater. Sol. Cells. 1999; **59**: 243–254. DOI: 10.1016/S0927-0248(99)00025-2.

[20] Yunaz IA, Hashizume K, Miyajima S, Yamada A, Konagai M. Fabrication of amorphous silicon carbide films using VHF-PECVD for triple-junction thin-film solar cell applications. Sol. Energy Mater. 2009; **93**: 1056–1061. DOI: 10.1016/j.solmat.2008.11.048.

[21] Lee CT, Tsai LH, Lin YH, Lin GR. A chemical vapor deposited silicon rich silicon carbide p–n junction based thin-film photovoltaic solar

cell. ECS J. Solid State Sci. Technol. 2012; **1**: Q144–Q148. DOI: 10.1149/2.005301jss.

[22] Dharmadasa IM. Third generation multi-layer tandem solar cells for achieving high conversion efficiencies. Sol. Energy Mater. 2005; **85**: 293–300. DOI: 10.1016/j.solmat.2004.08.008.

[23] Jiang CW, Green MA. Silicon quantum dot superlattices: modeling of energy bands, densities of states, and mobilities for silicon tandem solar cell applications. J. Appl. Phys. 2006; **99**: 114902. DOI: 10.1063/1.2203394.

[24] Lin GR, Lo TC, Tsai LH, Pai YH, Cheng CH, Wu CI, Wang PS. Finite silicon atom diffusion induced size limitation on self-assembled silicon quantum dots in silicon-rich silicon carbide. J. Electrochem. Soc. 2011; **159**: K35–K41. DOI: 10.1149/2.014202jes.

[25] Lo TC, Tsai LH, Cheng CH, Wang PS, Pai YH, Wu CI, Lin GR. Self-aggregated Si quantum dots in amorphous Si-rich SiC. J. Non-Cryst. Solids. 2012; **358**: 2126–2129. DOI: 10.1016/j.jnoncrysol.2012.01.013.

[26] Cheng Q, Tam E, Xu S, Ostrikov K. Si quantum dots embedded in an amorphous SiC matrix: nanophase control by non-equilibrium plasma hydrogenation. Nanoscale. 2010; **2**: 594–600. DOI: 10.1039/B9NR00371A.

[27] Kunii T, Honda T, Yoshida N, Nonomura S. Optical properties of microcrystalline 3C-SiC:H films measured by resonant photothermal bending spectroscopy. J. Non-Cryst. Solids. 2006; 352: 1196–1199. DOI: 10.1016/j.jnoncrysol.2006.01.074.

[28] Cheng CH, Lin YH, Chang JH, Wu CI, Lin GR. Semi-transparent Si-rich Si_xC_{1-x} p-i-n photovoltaic solar cell grown by hydrogen-free PECVD. RSC Advances. 2014; **4**: 18397–18405. DOI: 10.1039/C3RA41173G.

[29] Solomon I, Schmidt MP, Tran-Quoc H. Selective low-power plasma decomposition of silane-methane mixtures for the preparation of methylated amorphous silicon. Phys. Rev. B 1988; **38**: 9895–9901. DOI: 10.1103/PhysRevB.38.9895.

[30] Miyake S. Tribological properties of hard carbon films: extremely low friction mechanism of amorphous hydrogenated carbon films and amorphous hydrogenated SiC films in vacuum. Surf. Coat. Technol. 1992; 54/55: 563–569. DOI: 10.1016/S0257-8972(07)80083-1.

[31] Yu Z, Pereyra I, Carreno MNP Wide optical band gap window layers for solar cells. Sol. Energy Mater. Sol. Cells. 2001; **66**: 155–162. DOI: 10.1016/S0927-0248(00)00168-9.

[32] Kuhman D, Grammatica S, Jansen F. Properties of hydrogenated amorphous silicon carbide films prepared by plasma-enhanced chemical vapor deposition. Thin Solid Films 1989; **177**: 253–262. DOI: 10.1016/0040-6090(89)90573-7.

[33] Stutzmarin M, Jackson WB, Tsai CC. Light-induced metastable defects in hydrogenated amorphous silicon: a systematic study. Phys. Rev. B. 1985; **32**: 23–46. DOI: 10.1103/PhysRevB.32.23.

[34] Tawada Y, Okamoto H, Hamakawa Y. a-SiC:H/a-Si:H heterojunction solar cell having more than 7.1% conversion efficiency. Appl. Phys. Lett. 1981; **39**: 237–239. DOI: dx.doi.org/10.1063/1.92692.

[35] Abdelkrim M, Loulou M, Gharbi R, Fathallah M, Pirri CF, Tresso E. Static and dynamic electrical study of a-SiC:H based p-i-n structure, effect of hydrogen dilution of the intrinsic layer. Solid-State Electron. 2007; **51**: 159–163. DOI: 10.1016/j.sse.2006.11.015.

[36] Banerjee C, Narayanan KL, Haga K, Sritharathikhun J, Miyajima S, Yamad A, Konagai M. Fabrication of microcrystalline cubic silicon carbide/crystalline silicon heterojunction solar cell by hot wire chemical vapor deposition. Jpn. J. Appl. Phys. 2007; **46**: 1–6 DOI: 10.1143/JJAP.46.1.

[37] Irikawa J, Miyajima S, Watahiki T, Konagai M. High efficiency hydrogenated nanocrystalline cubic silicon carbide/crystalline silicon heterojunction solar cells using an optimized buffer layer. Appl. Phys. Express. 2011; **4**: 092301.

[38] Yunaz IA, Nagashima H, Hamashita D, Miyajima S, Konagai M. Wide-gap a-Si$_{1-x}$C$_x$:H solar cells with high light-induced stability for multijunction structure applications. Sol. Energy Mater. Sol. Cells. 2011; **95**: 107–110. DOI: 10.1016/j.solmat.2010.04.039.

5

Water Splitting Using GaN-based Working Electrodes for Hydrogen Generation with Bias by Solar Cells

Yen-Yu Chen

Department of Photonics, National Cheng Kung University,
Tainan 70101, Taiwan

Abstract

We demonstrated PEC hydrogen generation using GaN-based semi-conductors as the working electrode for water splitting under solar illumination. The generation rate of hydrogen could be improved by applying external bias onto the working electrodes to enhance the separation of photogenerated electron-hole pairs in the semiconductor and the charge transfer at the interface between semiconductor and electrolyte. Instead of using an external bias provided by power supply, a solar cell (SC) was used to raise the driving force to increase the rate of hydrogen production. The bandgap of GaN is suitable for photoeletrolysis, and it is resistance to the solution. In order to increase the absorption of solar spectrum, we fabricated hybrid working electrodes, such as n-GaN and InGaN. Prove that the hybrid working electrodes with SC to bias the PEC cells could improve the efficiency of hydrogen generation.

Keywords: PEC, water splitting, SC, GaN.

5.1 Introduction

Recently, fossil fuels have been a main source of energy around the world. Nowadays, the human have been using much fossil fuel to develop the economic and industrial. The greenhouse gases cause the global warming

Green Photonics and Smart Photonics, 87–96.

and severe climate changing. The more and more energy were consumed year by year. Therefore, the global are dedicated to develop the renewable energy such as solar cell (SC), biomass energy, wind energy and hydrogen energy in replace of fossil fuels like coal, petroleum at present. As result, hydrogen has been expected to be utilized as a new energy source. On the other hand, H_2 generation by water splitting using a photocatalyst is a clean method, because only water and solar light energy are consumed and no carbon dioxide is emitted. Not only can we develop economic and industrial, but also prevent the global warming so as to protect our environment at the same time. In 1972, Fujishima and Honda [1] used Titanium Dioxide as the working electrode to split water into hydrogen and oxygen under UV light illumination. Hydrogen can produce energy in a heat engine or provide electrical energy in a fuel cell via reacting with hydrogen. It is the simplest, abundant, available element which exists as a compound form in plant, hydrocarbons, and water. The use of sunlight to split water into hydrogen and oxygen via photoelectrolysis is a promising approach for the generation of renewable energy at present [2].

To achieve photocatalytic water splitting, the bandgap of the semiconductor need to straddle the reduction and oxidation potentials of water, which are +0 and +1.23 V versus NHE, respectively. For a working electrode made of n-type semiconductor, the holes drive the oxidation of water on the photoelectrode, and the electrons transport within semiconductor and in turn move to the cathode electrode leading to the reduction reaction [3]. When the semiconductor contact with electrolyte, the initial Fermi level in the electrolyte is arbitrarily drawn just above the O_2–H_2O redox level. After equilibrium in the dark, the Fermi level in the semiconductor equilibrates with the electrolyte Fermi level, producing a band bending. The Fermi level is below the H^+/H_2 potential. In order to raise the Fermi level in the metal counter electrode above the H^+/H_2 potential, an external photoanode bias must be applied. This bias also provides the overvoltage at the metal cathode to improve charge-separation. A smaller bandgap is desirable to utilize the wide range solar spectrum efficiently. Therefore, control of the energy level of the band edges is important for photocatalysts with high efficiency. Group III–nitride semiconductors are desirable as photocatalysts for two reasons. First, the band edge potentials of GaN and AlN straddle the water redox potentials (1.23 V) and therefore have the potential to split water without extra bias; second, they can absorb visible light using InGaN alloys, the bandgap energy of which can be easily controlled by changing the group-III content. Although GaN is potentially resistant to aqueous solution and its energy band is suitable for water photoeletrolysis, GaN only absorbs UV light which is about 5% of the solar spectrum. The bandgap energy of the $In_xGa_{1-x}N$ can be tuned from 0.7 to 3.4 eV by changing the indium content to fit the most of the terrestrial solar spectrum [4]. Therefore, InGaN-based working electrodes for water splitting are expected to be more efficient than the GaN working electrodes.

The working electrodes used in this study featuring meshed metal contacts with SiO_2 protection layer were immersed in electrolyte to enhance the collection efficiency of photogenerated carriers [5]. Although the external bias applied to working electrode could improve the rate of hydrogen generation, the external bias on PEC water splitting requires extra input power except light illumination.

Instead of using an external bias provided by power supply, we utilize GaN-based working electrodes with assisted bias supplied by the SC for PEC-splitting water under light illumination to increase the efficiency of hydrogen generation. The SC was connected in series between the InGaN-based working electrode and the Pt counter electrode to bias the PEC cell. In principle, the water splitting for generating hydrogen involves two mechanisms during the light illumination if the SC supplies bias voltage larger than 1.23 V. The mechanisms are photoelectrolysis and electrolysis of water splitting. To solely evaluate the efficiency of photoelectrolysis or electrolysis of water splitting, the output voltage of SC was varied through changing the type of SC to allow the PEC cell with bias less or larger than 1.23 V [6].

5.2 Experiments

The n-GaN epitaxial layer was grown on the sapphire (Al_2O_3) substrate by MOVPE. The low-temperature GaN nucleation layer with a thickness of 30 nm was deposited on the sapphire substrate followed by the 2-μm thick unintentionally doped GaN (u-GaN) buffer layer. The n-InGaN epitaxial layer was grown on the other side of double-polished sapphire substrate by MOVPE. We fabricated the different hybrid working electrodes. The light source was equipped a 300-W Xenon lamp with an AM1.5G filter to serve as under solar light illumination. A Pt wire was used as the counterelectrode, and 1 M NaCl was used as electrolyte. In addition, we used potentiostat (Autolab-PGSTAT128N) to supply the external bias and measure the PC during the PEC water-splitting process.

The experiment setup of water splitting and photoelectrolysis via the n-GaN photoanode with assisted bias generated from the SC are shown in

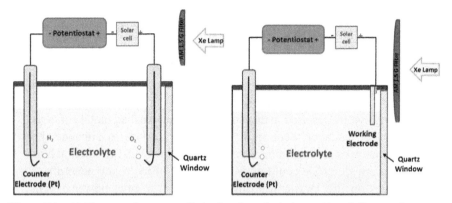

Figure 5.1   (a) The setup for water splitting by electrolysis (setup A) and (b) setup for water splitting by photoelectrolysis and electrolysis (setup B).

Figure 5.1. The positive and the negative electrodes of the SC were connected to the n-GaN working electrode and the potentiostat, respectively. In order to compare the efficiency of applying external bias by SC, we also connect potentiostat in series with the working electrode and the counter electrode to measure the PC density without extra bias by SC. The incident light from a Xenon lamp passed through an AM1.5G filter and illuminated on the SC. Connected the SC and the potentiostat (Autolab) to measure the typical I–V curve under illumination. The incident light from a Xenon lamp passed through an AM1.5G filter and illuminated on the SC. Connected the SC and the potentiostat (Autolab) to measure the typical I–V curve under illumination, as shown in Figure 5.2. In this study, GaN-based semiconductors associated with metal contacts were served as working electrodes without external bias to conduct the PEC water-splitting process under the light illumination.

5.3 Results and Discussions

Figure 5.1a, illustrates a water splitting setup for hydrogen generation by electrolysis powered by SC providing an output voltage of 2.5 V. Both the working electrode and the counter electrode are Pt. In order to prove the efficiency of photoelectrolysis and water splitting is better than the just of water splitting. The results indicated that the hybrid working electrodes exhibited a marked improvement in the efficiency of hydrogen generation. The output voltage of potentiostst was set to be zero. To evaluate the effect of

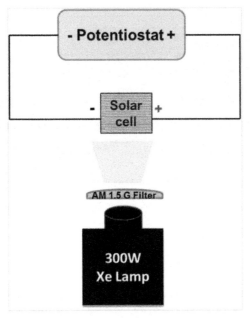

Figure 5.2 The experimental setup of SC *I–V* curves.

water splitting by photoelectrolysis on hydrogen generation, the Pt electrode was replaced by the GaN-based working electrode, as shown in Figure 5.1b. Notably, the PC in the setup with GaN electrode was markedly higher than the reference setup without GaN electrode. The enhancement of PC in setup Figure 5.1b could be attributed to additional effect of photoelectrolysis originated from the GaN electrode. However, the PCs observed from the setup with GaN electrode decreased significantly with reaction time, as shown in Figure 5.3. This result could be attributed to the fact that the n-GaN film was increasingly etched away during the PEC process.

To further improve the PC in the PEC cell, different SCs was incorporated into the hybrid PEC cells to evaluate the optimum operation point. The measured PC were marked with different colors symbols on the *I–V* curve of SC. It has been noted that the marked symbols denote the operating points of the hybrid system. The effective resistance of the PEC cell could be tuned to match the operation point on P_{max} of the SC.

The experiment setup of photoelectrolysis via the n-GaN photoanode biased 0.5 V by SC shows less PC. The setup of photoelectrolysis via the n-GaN photoanode biased 0.8 V by GaAs SC shows larger PC.

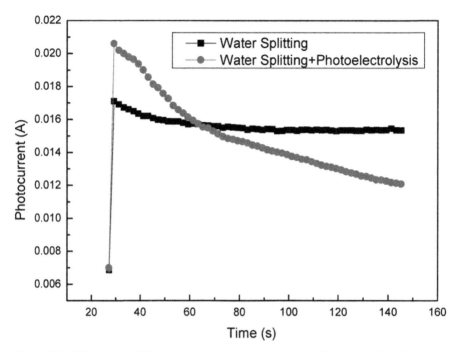

Figure 5.3 PC of water splitting and photoelectrolysis via the n-GaN photoanode with assisted bias generated from the 2.5 V SC.

As shown in Figure 5.4, the effective resistance of the PEC cell could be tuned to match the operation point on P_{max} of the SC. In order to improve the operation point to match the maximum power, we used two GaAs and a 0.5-V SCs in series to increase the bias on hybrid PEC cells. Figure 5.5 shows the PC for water splitting by electrolysis and photoelectrolysis. The setup for water splitting by electrolysis used two Pt electrodes connecting in series two GaAs and a Si SCs with output voltage of 2 V, as shown in Figure 5.1a; the setup for water splitting by photoelectrolysis and electrolysis used the n-GaN photoanode connected in series two GaAs and a Si SCs with output voltage of 2 V, as shown in Figure 5.1b. As one can see that the PEC cell with n-GaN photoanode exhibited higher PC. This result could be attributed to the fact that n-GaN photoanode could contribute additional PC to the PEC cell due to the light absorption by the n-GaN to generate e-h pairs.

In order to increase PC, we demonstrated that the hybrid working electrodes. The structure of n-GaN could absorb light at the short wavelength, then

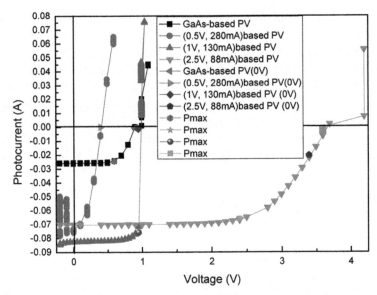

Figure 5.4 The measured PCs with different SC were marked with distinctive color symbols on the *I–V* curve of different SC.

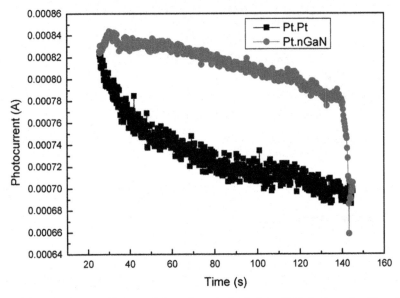

Figure 5.5 PC of water splitting and photoelectrolysis via the n-GaN photoanode with assisted bias generated from the two GaAs and 0.5 V SCs in series of connection.

n-InGaN could absorb light at the long wavelength, as shown in Figure 5.6. This way could decrease the possibility of carrier recombination and increase the reaction paths. In Figure 5.7a, the measured PC without bias by SC shows the less PC density. However, the enhancement of PC density in Figure 5.7b could be attributed to external bias by SC 2.5 V. The hybrid PEC cells with bias provided by SCs on working electrode could increase efficiency of hydrogen generation. Figure 5.8 shows the incident IPCE of hybrid working electrode is larger than the simple layer. The hybrid working electrode could absorb more incident light than the simple layer to utilize most of the solar spectrum efficiently.

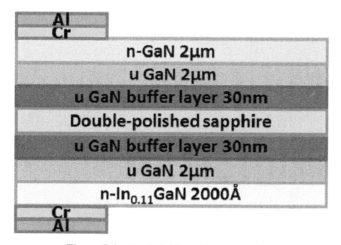

Figure 5.6   The hybrid working electrodes.

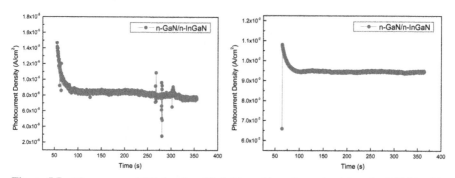

Figure 5.7   The measured PC density of hybrid working electrodes (a) under 0 V (b) with bias generated from the 2.5 V SC.

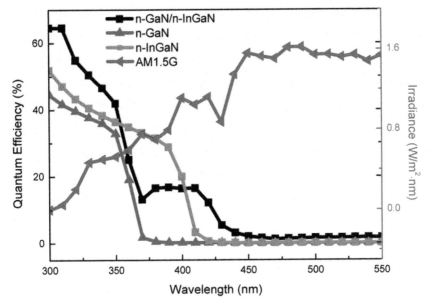

Figure 5.8 Incident IPCE of working electrodes.

5.4 Conclusion

We have demonstrated that the PEC cells with external bias provided by SCs on working electrode could increase efficiency of hydrogen generation. Prove the PC of photoelectrolysis and water splitting is better than just water splitting. Connect different SC to the working electrode to measure PC of the PEC cells under light illumination, we could tune the operation point to fit the P_{max} point for increase the utility of SC in the PEC cells. The Nitride-based hybrid working electrodes could absorb more light energy to improve the efficiency of hydrogen generation.

References

[1] A. Fujishima, "Electrochemical photolysis of water at a semiconductor electrode," Nature, vol. 238, pp. 37–38, 1972.
[2] A. NOZIK, "Electrode materials for photoelectrochemical devices," Journal of Crystal Growth, vol. 39, pp. 200–209, 1977.
[3] K. K. a. K. H. Akira Fujishima, "Hydrogen production under sunlight with an electrochemical photocell," J. Electrochem. Soc., vol. 122, pp. 1487–1489, 1975.

[4] I. Waki, D. Cohen, R. Lal, U. Mishra, S. P. DenBaars, and S. Nakamura, "Direct water photoelectrolysis with patterned n-GaN," *Applied Physics Letters*, vol. 91, p. 093519, 2007.

[5] S.-Y. Liu, J. K. Sheu, C.-K. Tseng, J.-C. Ye, K. H. Chang, M. L. Lee, *et al.*, "Improved Hydrogen Gas Generation Rate of n-GaN Photoelectrode with SiO[sub 2] Protection Layer on the Ohmic Contacts from the Electrolyte," *Journal of The Electrochemical Society*, vol. 157, p. B266, 2010.

[6] K. Fujii, T. Karasawa, and K. Ohkawa, "Hydrogen Gas Generation by Splitting Aqueous Water Using n-Type GaN Photoelectrode with Anodic Oxidation," *Japanese Journal of Applied Physics*, vol. 44, pp. L543–L545, 2005.

6

Fiber Amplifiers for Photonic Communication and Sensing

Yi-Lin Yu[1], Shien-Kuei Liaw[1] and Jingo Chen[2]

[1]National Taiwan University of Science and Technology, Taiwan
[2]National Center for Research on Earthquake Engineering, Taiwan

Abstract

In this chapter, several types of fiber amplifier have been demonstrated and proposed, which could be used in communication and sensing, including Erbium doped fiber amplifier (EDFA), Raman fiber amplifier (RFA), hybrid fiber amplifier (EDFA + RFA) and high power fiber amplifier. The authors also used the concept of residual pumping reused to optimize the gain efficiency of C + L band signals.

Keywords: EDFA, EYDF, RFA.

6.1 Introduction

The rapid growth of data traffic in optical communications and optical networks demands extensive research into wideband optical amplifiers as the roles they play have become increasingly important [1, 2]. So far, there are many methods to amplify the C + L band signals such as DFA, Raman fiber amplifier (RFA), and SOA. DFAs are optical amplifiers that used a doped optical fiber as a gain medium to amplify optical signals. The signals and pumping light source are multiplexed into the doped fiber, and the signal is amplified through interaction with the doping ions. The most common example is the EDFA, where the core of a silica fiber is doped with trivalent Erbium ions and can be efficiently pumped with laser light source at a wavelength of 980 or 1480 nm, and exhibits gain in the C + L band region. SOAs are amplifiers which use a semiconductor to provide the gain medium. These amplifiers

Green Photonics and Smart Photonics, 97–114.
© 2016 *River Publishers. All rights reserved.*

have a similar structure to that of Fabry–Perot LDs but with antireflection design elements at the end faces. In a RFA, the signal is intensified by Raman amplification. Unlike the EDFA and SOA, the amplification effect of RFA is achieved by a non-linear interaction between the signal and a pump laser within an optical fiber [3–5]. For the C-band, EDFA is a mature and widely used technology, owning to its higher gain and lower noise figure (NF) than other methods. RFA is a convenient method to amplify various wavebands, and it has even lower NF than that of EDFA. However, both EDFA and RFA are easier to connect to fiber network than SOA due to the fiber gain medium [62]. In addition, fiber amplifier is a broad band light source which can be used for smart sensing [7, 8]. In this chapter, we proposed four kinds of fiber amplifier, EDFA, RFA, hybrid fiber amplifier, and high-power fiber amplifier. In the first and third schemes, we proposed a bidirectional EDFA and a bidirectional hybrid fiber amplifier. Both the C- and L-band signals were amplified by using a single wavelength pump LD. Moreover, we employed the concept of residual pumping power reuse to optimize the gain efficiency of C + L band signals. In the second scheme, we proposed a new architecture of RFA based on signal and pump double-pass the gain medium and cascaded FBGs to increase the pumping efficiency. In the final scheme, we proposed a high-power fiber amplifier by using EYDF as the gain medium. EYDF attracts increased interest due to the possibility of doping fiber with high gain, pump efficiency, and broadening of the absorption band from 850 to 1100 nm which offers greater flexibility in selection of the pump wavelength.

6.2 Experimental Setup

6.2.1 Bidirectional EDFA

Figure 6.2a depicts the proposed bidirectional C + L band EDFAs sharing the same pump LD. The C-band EDFA is for downstream signal amplification at the upper path and the L-band EDFA is for upstream signal amplification at the lower path. The C-band EDFA consists of a wavelength division multiplexer (WDM), a 3-m long EDF, and a pump reflector. The EDF_1 (DF1500L-980) absorption coefficients are 12.4 dB/m@979 nm and 18.79 dB/m@1531 nm, respectively. A FBG based pump reflector with 20 dB reflectivity is inserted in the C-band path, and its spectrum is shown in Figure 6.2b. Since the pumping power will not be absorbed completely, we insert a pump reflector into the C-band EDFA. The residual pumping power can be reabsorbed and reused by the EDF_1 after it loops back to the EDF_1 using the FBG-based pump

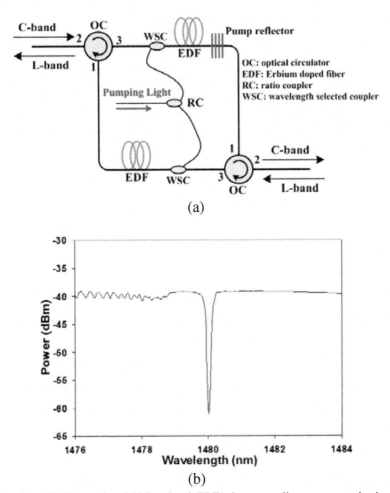

Figure 6.1 (a) Schematic of bidirectional EDFA in a recycling pump mechanism and (b) Transmission spectrum of FBG at 1480 nm.

reflector. The L-band EDFA is constructed using a WDM and 10 m long EDF. The EDF$_2$ (RightWave^{TM} EDF LRL Reduced Length) absorption coefficients are 10 dB/m@1480 nm and 33.5 dB/m@1530 nm. The left-hand-side optical circulator (OC) is used to add the downstream signals to the C-band EDFA and to receive the upstream signals from the L-band EDFA. Conversely, the right-hand-side OC is used to add the upstream signals to the L-band EDFA and to receive the downstream signals from the C-band EDFA. The pumping power of 200 mW uses only one pump LD. We assign 60 mW of pumping power

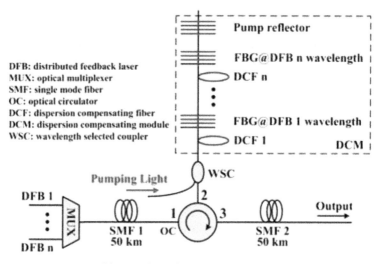

Figure 6.2   Configuration of the proposed RFA [12].

to the C-band EDFA. A residual pumping power of 14.6 mW is available for the C-band EDFA after 60 mW pumping power passing through the C-band EDFA. It means only 45.4 mW or less of the pump power used for the first round pumping. The 14.6 mW residual pump power corresponding to 1/3 of the original pump power is used for the second round pumping, thanks to the FBG-based pump reflector in the C-band EDFA. For the L-band EDFA, the residual pumping power is negligible after it passes through the EDFA and so it is not necessary to use a FBG-based pump reflector.

6.2.2 Raman Fiber Amplifier

In a RFA-based WDM system, it is difficult to use only one segment of DCF for compensating WDM channels in the whole C and/or L-band. It is attributed to the mismatch of dispersion slope of the DCF and the standard fiber, that is, SMF. In order to solve this problem, several FBGs could be embedded in the DCF. In this work, each signal travel path in the DCF is controlled by an FBG, whose central wavelength is designed to match the signal wavelength. Each segment of the DCF's length can be predicted on the basis of the following equation:

$$L_{\mathrm{DCF}} = -\frac{L_{\mathrm{SMF}} D_{\mathrm{SMF}}(\lambda)}{2 D_{\mathrm{DCF}}(\lambda)} \tag{6.1}$$

where $D_{\text{SMF}}(\lambda)$ and $D_{\text{DCF}}(\lambda)$ are the dispersion parameter of SMF and DCF, respectively, and L_{SMF} is the length of SMF. Because the WDM signals pass through the DCF twice, there is a 1/2 factor in this expression path of the round-trip design. Figure 6.1 depicts the configuration of our proposed RFA. The WDM signals are fed into a 50-km SMF via a WDM multiplexer and then travel through port 1 to port 2 of the optical circulator (OC). Raman pump and signals are multiplexed into the dispersion compensation module (DCM) via a Raman coupler. The DCM is composed of several FBGs, several segments of DCF, and a FBG-based pump reflector. Each FBG is matched with a certain WDM channel. Inside the DCM, different signals travel through different lengths of DCF. For example, signal 1 passes through only the DCF 1 and then is reflected by the FBG 1, while signal 2 passes through both the DCF 1 and DCF 2 and then is reflected only by the FBG 2 and so forth. The length of each segment of the DCF is determined by using Equation (6.1) to eliminate the residual dispersion of WDM signal channels. The Raman pump passes through the whole DCF and FBGs in the DCM firstly, and then the residual pump power comes back from pump reflector again. Thus, the pump power double-passes the gain medium of the DCF for increasing the pumping efficiency. In the double-pass scheme RFA, the forward and backward power evolution of pump signals can be expressed in terms of the following equation [9]:

$$\frac{\mathrm{d}P^{\pm}(z, v_i)}{\mathrm{d}z} = \mp \alpha(v_i) P^{\pm}(z, v_i) \pm P^{\pm}(z, v_i)$$

$$\sum_{m=1}^{i-1} \frac{g_{\text{R}}(v_m - v_i)}{\Gamma A_{\text{eff}}} \left[P^{\pm}(z, v_i) + P^{\mp}(z, v_i) \right] \mp P^{\pm}(z, v_i)$$

$$\sum_{m=i+1}^{n} \frac{v_i}{v_m} \frac{g_{\text{R}}(v_i - v_m)}{\Gamma A_{\text{eff}}} \left[P^{\pm}(z, v_i) + P^{\mp}(z, v_i) \right], \qquad (6.2)$$

where $P^+(z, v_i)$ and $P^-(z, v_i)$ are optical power of the forward and the backward propagating waves within infinitesimal bandwidth around v_i, respectively. $\alpha(v_i)$ is the attenuation coefficient of the corresponding wavelength v_i. A_{eff} is the effective area of the optical fiber; $g_{\text{R}}(v_i - v_m)$ is the Raman gain parameter at frequency v_i due to pump at frequency v_m; the factor Γ accounts for polarization randomization effect, which value lies between 1 and 2. Besides dispersion compensation, we may also carry out the gain equalization by adjusting the reflectivity of each FBG. Here, the objective function is

$$f_i(R_i) = \text{abs}\left(\frac{P_i - \text{avg}[P(R)]}{\text{avg}[P(R)]} \right) \qquad i \in [i, N] \qquad (6.3)$$

where $P_i(R)$ is the signal power of the ith WDM at FBG with reflectivity R_i, and our aim is to minimize the objective function to zero. Considering a possibly large channel number in a real WDM system, we use the Broyden method [10, 11] as it has proved highly effective in solving non-linear systems equations. It is globally convergent and provides an easier approximation to the Jacobian matrix for zero finding. The proposed DCF module is a non-linear system as proper dispersion compensation and FBG reflectivity have to be determined.

6.2.3 Bidirectional Hybrid Fiber Amplifier

The Raman shift concept is illustrated in Figure 6.3. For a pump source at 1495 nm for example, the corresponding gain peak can be described using the following equation:

$$\triangle \lambda = -\lambda \frac{\Delta f}{f} = -\lambda^2 \frac{\Delta f}{c} = 94.9 \text{ nm} \tag{6.4}$$

where $\Delta f = -13$ THz and $\triangle \lambda = 94.9$ nm are the total amount of detuning with respect to the pump frequency and wavelength, respectively. The maximum gain occurs at 1589.5 nm in the L-band region. The C-band EDFA can also use the same 1495 nm pump LD. The gain is only a little smaller than the one achieved using 1480 nm. Thus, a single-wavelength pump LD may be used as the pump for both the C-band EDFA and L-band RFA simultaneously. Figure 6.3a shows the proposed bidirectional hybrid, C + L-band EDFA/RFA

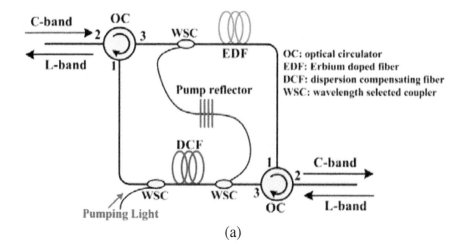

(a)

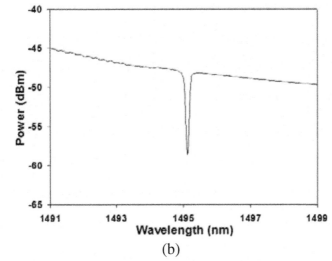

(b)

Figure 6.3 (a) Schematic of bidirection hybrid fiber amplifier using pump sharing concept (b) Transmission spectrum of FBG at 1495 nm.

for WDM applications. The C-band EDFA is for the downstream signal amplification. The L-band RFA for the upstream signal amplification consists of a length of DCF. Instead of using a high-power pump LD, we used two single-wavelength pump LDs at 1495 nm to launch power into the RFA via a PBC, which is used as depolarizer to reduce the PDG effect of EDFA and RFA. The residual pump power was then routed to the C-band EDF via a pair of WDM couplers. Since the residual pump power may still be too much for the EDFA, a ratio pump reflector was placed between the WDM pair to reflect part of the residual pump back to the L-band RFA for further pumping, and its spectrum is shown in Figure 6.3b. The left hand side OC is used to add the C-band signals to the EDFA as well as receive the L-band signals from the RFA. The right-hand-side OC is used to add the L-band signals to the RFA as well as receive the C-band signals from the EDFA. The signals for both bands have unidirectional paths, thanks to the interport isolation of the OCs. Note that the DCF plays a dual role as both the gain medium and dispersion compensator for the RFA.

6.2.4 High-Power Erbium Ytterbium Co-Doped Fiber Amplifier (EYDFA)

Figure 6.4 shows the two-stage high-power fiber amplifier configuration by using EYDF as the gain medium. In the EYDF, energy transfer from the excited

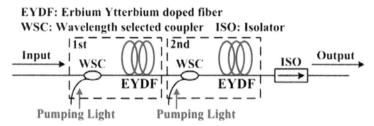

Figure 6.4   Two-stage high-power fiber amplifier with double forward scheme.

state of Yb to that of Er is utilized to form population inversion between lasing levels of Er, and the signal is amplified through stimulated emission. The limitation of pump current here we used could up to 8 A. We used these two isolators (ISO) for preventing the high-power backward ASE to damage the DFB laser. In the experimental setup, the WDM coupler we used was isolating wavelength division multiplexer (CP-IWDM). It could allow signal passes with only one direction, so it was unnecessary inserting an isolator between 1st stage and 2nd stage.

6.3 Results and Discussion

6.3.1 Bidirectional EDFA

Three input power conditions of −20/−15/−10 dBm are set for each channel into the C + L-band EDFA. Seven WDM channels are measured stepbystep, in 5 nm spacing, from 1530 to 1560 nm for the C-band EDFA. Similarly, seven WDM channels are measured step by step, at 5-nm spacing from 1570–1610 nm for the L-band EDFA. Figure 6.8a, b show the gain profiles of the C-band and L-band EDFAs, respectively, for different launched powers (P_{in}). For −10 dBm launched power, the average gains are 15.85 dB for the C-band EDFA and 15.54 dB for the L-band EDFA, respectively. Figure 6.9a, b show the NF characteristics of the C-band and L-band EDFAs, respectively, for different launched power levels (P_{in}). The average NFs are 5.19 dB for the C-band EDFA and 5.52 dB for the L band EDFA, respectively.

6.3.2 Raman Fiber Amplifier

The feasibility of configuration shown in Figure 6.1 is carried out by Optisystem 6.0. Without loss of generality, there are eight WDM channels inside the C-band, staring from 1530.8 to 1553.2 nm with channel spacing based on

ITU grid of 400 GHz (3.2 nm), with 0 dBm as the launched power level for each channel. Here, we would rather demonstrate the broadband RFA ability than crosstalk issue in a dense WDM system. The central pump wavelength is at 1451 nm with a pump power of 333 mW. In practice, this pump power level can be realized by combining two orthogonal, polarized pump lasers to obtain total higher power using a polarization beam combiner (PBC). As referred to [13], the signal absorption coefficients may be 0.2 and 0.3 dB/km for SMF and DCF, respectively. And the pump absorption in DCF is 0.55 dB/km. The Raman gain coefficient and the effective area of the DCF are assumed to be identical with those in Dianov [14]. First, we calculate the dispersion and determine the length of each DCF segment. In our simulation, the dispersion of SMF is 17 ps/nm/km at 1550 nm with a dispersion slope of 0.058 ps/km/nm^2; and the dispersion of DCF is −95 ps/nm/km at 1550 nm with a dispersion slope of −0.62 ps/km/nm^2. In a conventional configuration, the dispersion compensation is carried out by using only one segment of the DCF, and the transmission bandwidth or transmission speed is surely restricted. As shown in Figure 6.5, if we want to minimize the system residual dispersion in the conventional configuration, the length of the DCF should be equal to 9.1 km. In this condition, the maximum absolution of residual dispersion, which appears in both the longest and the shortest wavelengths, is +62 ps/nm and −62 ps/nm, respectively. So, the maximum transmission speed Rb is limited to 23 Gbit/s, as could be predicted by the following equation [15]:

$$R_{\mathrm{b}} = \sqrt{\frac{C}{4\,|D_{\mathrm{res}}|\,\lambda^2}} \tag{6.5}$$

where C is the speed of light in vacuum, D_{res} is the residual dispersion, and λ is the central wavelength.

By using DCM composed of FBGs among several DCF segments, the residual dispersion shown in Figure 6.5 can be theoretically eliminated. Besides the common DCF of 8.86 km in length, the calculated extra length of the DCF required is 0, 87, 91, 95, 99, 103, 108 and 113 m, for WDM channel of 1553.2, 1550.0, 1546.8, 1543.6, 1540.4, 1537.2, 1534.0, and 1530.8 nm, respectively. For the study of gain equalization, the reflectivity of each FBG is assumed to be 99.9% in the beginning. Note that the FBGs inside the DCM are uniform gratings rather than chirped FBGs. It is the DCF that single-handedly deals with the chromatic dispersion issue. As shown in Figure 6.6a, the solid curve depicts the output power when the FBGs are written at the correct positions for optimum dispersion compensation. The maximum power

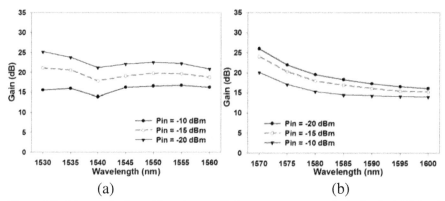

Figure 6.5 Measured gain profiles using tunable laser sources, respectively, for the (a) C-band EDFA and (b) L-band EDFA.

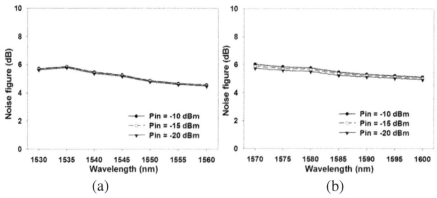

Figure 6.6 Measured NF profiles using tunable laser sources, respectively, for the (a) C-band EDFA and (b) L-band EDFA.

variation among channels is nearly 6 dB and the lowest output power is at the shortest wavelength. In order to equalize the output power, the FBG reflectivity corresponding to the lowest signal is set to 100% and other reflection ratios of FBGs are optimized using the Broyden method. As shown in Figure 6.6, when the reflectivity of FBGs is designed to 19.67, 19.1, 21.34, 26.45, 33.37, 44.93, 64.58, and 100%, respectively, the eight WDM channels are power-equalized simultaneously. In our lab, the home-made FBG's reflection ratio can be precisely controlled within ±5% accuracy easily. The output power of all channels could fall into the shadow region shown in Figure 6.6a, which indicates that the maximum output variation is less than ±0.5 dB as expected.

Note that the flattened amplification bandwidth is as large as 23 nm. Even so, only a single-pump laser is used in this study. In this case, the splicing loss of FBG–DCF junction and sideband loss of FBG are assumed negligible for simplification. On the other hand, if we assume that the absorption coefficient of the DCF increases to 0.5 dB/km as with some commercial products, the required pump power will increase a little bit to 383 mW for obtaining the same net gain. The reflectivity of FBGs then is modified as 21.3, 19.8, 21.4, 26.0, 32.3, 43.3, 62.8, and 100% for the corresponding channels. In our recent work, we successfully proved that the simulated result [9] for RFA agrees well with that of the experimental work [16]. So, the simulated performance is realistically achievable because a similar technique is applied here. To confirm the system feasibility, a comprehensive numerical simulation of the signal transmission characteristics in light wave system employing this RFA is evaluated. Two configurations, with and without residual dispersion compensation, are employed in our simulation for comparison. A pseudo-random-binary-sequence (PRBS 10^{23}–1 NRZ formats is applied to an intensity modulation of WDM channels in 40 Gb/s speed for each channel. The total signal envelope propagates through the fiber span including 100 km SMF is modeled by the modified non-linear Schrödinger Equation (NLSE) [17]

$$
i\frac{\partial A_j}{\partial z} + \frac{i}{v_{gj}}\frac{\partial A_j}{\partial t} - \frac{1}{2}\beta_{2j}\frac{\partial^2 A_j}{\partial t^2} + \frac{i}{6}\beta_{3j}\frac{\partial^3 A_j}{\partial t^3}
$$
$$
+ \gamma\left(|A_j|^2 + 2\sum_{m \neq j}^{M}|A_m^2|\right)A_j = \frac{i\alpha}{2}A_i \qquad (6.6)
$$

where v_{gj} is group velocity of jth channel, β_{2j} is GVD parameter, β_{3j} is TOD parameter, γ is nonlinear coefficient, and α accounts the loss. We incorporate the signal/pump double-pass Raman amplification effect by adding the distributed RFA gain coefficient into the NLSE [18]. The calculated equivalent fiber loss for the first channel is shown in Figure 6.6(b) and the DCF's loss coefficient is also plotted. The RFA in simulation can provide 20-dB of net gain. Then we compare two different schemes of residual dispersion compensated and without residual dispersion compensation. For example, one may write the FBGs at the same position for the latter case. In both schemes, the total loss attributed by 100 km SMF is compensated completely by the RFA. The BER with respect to the input signal wavelength is shown in Figure 6.7(a). The error-free condition (BER $\leq 10^{-11}$) could be obtained as NRZ data format is used. On the other hand, without residual dispersion, compensation will lead to worse system performance both for the shortest and

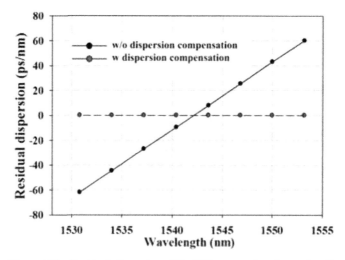

Figure 6.7   Residual dispersion of the RFA versus signal wavelength.

the longest wavelengths. In our simulation, only three kinds of impairments such as amplifier accompany noise, residual dispersion, and non-linearity are considered to be the overall impacts to close the eye diagram. In such a condition, the noise will be a Gaussian distribution [19]. So, Equation 6.7 could be used to convert Q value to BER directly.

$$\mathrm{BER} = \frac{1}{2}\mathrm{erfc}\left(\frac{Q}{\sqrt{2}}\right) \tag{6.7}$$

where erfc is the error function. As shown is Figure 6.7b, the Q values become larger than 6.6 dB for all WDM channels when the proposed method to compensate the residual dispersion is used. Otherwise, it will lead to as large as 2 dB degradation in Q value, corresponding to BER degradation from 10^{-12} to 10^{-7} for both the longest and the shortest wavelengths. These results confirm the feasibility of RFA which we proposed.

6.3.3 Bidirectional Hybrid Fiber Amplifier

To measure the gain and NF characteristics, two 1500–1600 nm tunable laser sources were used for both bands. Seven WDM channels were measured step by step, in 5 nm spacing, from 1530–1565 nm for the EDFA, and nine WDM channels were measured step by step, in 5 nm spacing from 1570–1610 nm for the RFA. The total pump power of 500 mW was achieved using two

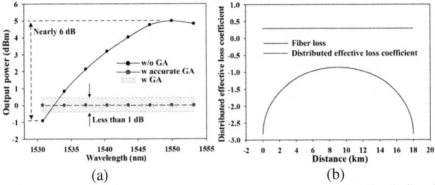

Figure 6.8 (a) Output power of the RFA versus signal wavelength and (b) distributed effective loss coefficient of signal 1 along the DCF.

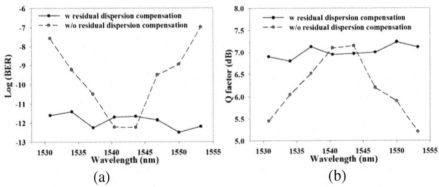

Figure 6.9 (a) BER as a function of input signal wavelength and (b) Q value as a function of input signal wavelength.

identical wavelengths pump LDs. The length was 3 m and the absorption was 11 dB/m@1480 nm for the EDFA. The fiber loss, dispersion, and dispersion slope for the DCF are 0.4 dB/km, –95 ps/nm/km and –0.62 ps/nm$_2$/km, respectively. Three power conditions of –20/–15/–10 dBm were set for each channel for launching into the hybrid EDFA/RFA. A residual pump power of 54.88 mW was available for the EDFA as well as providing recycled pumping for the RFA. After trial and error, the optimum reflectivity of a "home-made" pump reflector was determined with 38% reflectance. Figures 6.10a, b show the gain profiles of the EDFA and RFA, respectively, for different launched power (P_{in}) levels. The average gains are 11.52 dB for the C-band EDFA and 11.72 dB for the longer band of L-band RFA, respectively, for –10 dBm

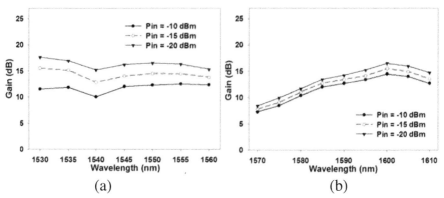

Figure 6.10   Measured gain profiles using a tunable laser source, respectively, for (a) C-band EDFA gain and (b) L-band RFA gain.

launched power for each channel. The power variation between C- and L-bands could be smaller if the pump reflector is precisely controlled at 43% in reflectance. Figures 6.11a, b show the NF characteristics of the EDFA and RFA, respectively, for different launched power levels (P_{in}). The average NF is 5.76 dB for the C-band EDFA and 4.28 dB for the L-band RFA, respectively.

6.3.4 High-Power EYDF Amplifier

To implement the gain optimization in the experiment of high-power fiber amplifier, we used a different length of EYDF, and the results were shown in Figure 6.12. In Table 6.1, we sorted the results from the Figure 6.2, and it was

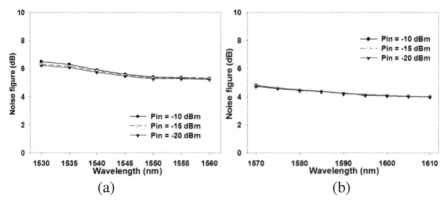

Figure 6.11   Measured NF profiles using a tunable laser source, respectively, for (a) C-band EDFA NF and (b) L-band RFA NF.

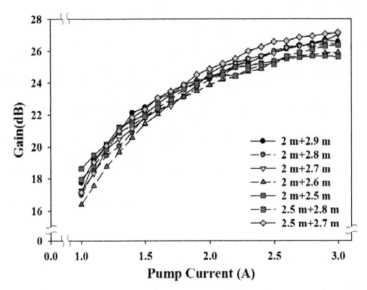

Figure 6.12 The gain profiles with proposed fiber amplifier with difference length of EYDF.

Table 6.1 The results with difference parameters of the high-power fiber amplifier

EYDF Length (1st, m)	EYDF Length (2nd, m)	Gain (dB)	Total Power (dBm)
2	2.9 m	26.58	26.82
2	2.8 m	26.43	26.64
2	2.7	27.03	27.31
2	2.6	25.91	26.81
2	2.5	25.65	26.48
2.5	2.8	26.38	27.67
2.5	2.7	27.15	27.61

Bias current@3A

obvious that the result with 2.5 m (1st) and 2.7 m (2nd) EYDF length has the maximum gain and output power, respectively.

To verify the high-power fiber amplifier operating, we increased the bias current of pump LD from 3- to 6 A with the same configuration. The results were shown in Figure 6.13. When the bias current was up to 6 A, the signal power, total power, and noise figure we measured were 32.42, 32.7 and 5.7 dB, respectively. The transfer efficiency was about 25.34% with 6 A bias current. In addition, polarization-dependent gain (PDG) is an important parameter of fiber amplifier. In this section, the PDG of high-power EYDFA we proposed was less than 0.23 dB.

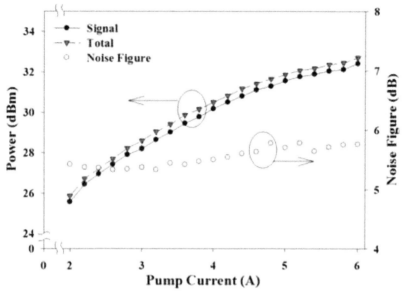

Figure 6.13   The measured signal power, total power, and noise figure.

6.4 Conclusion

Four kinds of fiber amplifiers, either Raman fiber amplifier, hybrid fiber amplifer or high power rare doped fiber amplifier were investigated and demonstrated. In first scheme, We proposed a RFA with dispersion and gain equlization characteristics based on FBGs. In order to compensate the dispersion slope mismatch between the transmission fiber and DCF, FBGs embedded in the different positions are used to control the travel length of each WDM signal. In second and third schemes, a single-wavelength pump source was shared between the C-band and L-band fiber amplifier. Gain equalization between these two bands can be further improved by adjusting the pump reflector reflectivity. In order to compensate the dispersion slope mismatch between the transmission fiber and DCF, FBGs embedded in the different positions are used to control the travel length of each WDM signal. Finally, we used Er/Yb codoped fiber as the gain medium of high power fiber amplifier. Up to 32 dBm output power was achieved in our experiment. All of the fiber amplifiers we proposed may find vast application in WDM system.

References

[1] Y. Sun, Y., J. W. Sulhoff, A. K. Srivastava, J. L. Zyskind, T. A. Strasser, J. R. Pedrazzani, C. Wolf, J. Zhou, J. B. Judkins, R. P. Espindola and A. M. Vengsarkar, "80 nm ultra-wide-band erbium-doped silica fiber amplifier," Electronics Letter, vol. 33, 1965–1967, 1997.

[2] H. Y. Hsu, Y. Lin Yu, Shien-Kuei Liaw, Ren-Yang Liu and Chow-Shing Shin, "Theoretical and experimental study of multifunction C+L band hybrid fiber amplifiers," Optics & Laser Technology, vol. 56, pp. 307–312, 2014.

[3] S. Sapre, and N. Gupta, "Review of EDFA gain performance in C and L band," International Journal of Innovation and Applied Studies, vol. 12, pp. 559–563, 2015.

[4] Q. Wang, C. Tian, Y. Y. Wang, and P. Wang, "Review of radiation hardening techniques for EDFAs in space environment," Proceedings of SPIE, vol. 9521, 2015.

[5] J. Gong, Y. Zhao, X. Zuo, X. Jiang, and C. Li, "Numerical study of Raman fiber amplifier based on cascading As-S fiber and As-Se fiber," Optical Review, vol. 22, pp. 483–488, 2015.

[6] X. Zhou, C. Lu, P. Shum, and T. H. Cheng, "A simplified model and optimal design of a multi-wavelength backward-pumped fiber Raman amplifier," IEEE Photonics Technology Letters, vol. 13, pp. 945–947, 2001.

[7] P. Lesiak, M. Szelag, M. Kuczkowski, M. Bieda, A. W. Domanski, and T. R. Wolinski, "Wavelength demodulation system for embedded FBG sensors using a highly birefringent fiber," Proceedings of SPIE, vol. 9441, 2014.

[8] J. Ono, T. Endo, K. Ohta, H. Ono, Y. Maeda, K. Senda, O. Koyama, and M. Yamada, "Broadband light source and Its application to near-infrared spectroscopy," Sensors and Materials, vol. 27, pp. 413–423, 2015.

[9] L. Dou, S. K. Liaw, M. Li, Y. T. Lin, and A. Xu, "Parameters optimization of high efficiency discrete Raman fiber amplifier by using the coupled steady-state equations," Optics Communications, vol. 273, pp. 149–152, 2007.

[10] C. G. Broyden, "A class of methods for solving nonlinear simultaneous equations," Mathematics of Computation, vol. 19, pp. 577–593, 1965.

[11] W. H. Press, S. A. Teuko, W. T. Vetterling, and B. P. Flannery, *Numerical Recipes in C: the art of scientific computing*, Cambridge University Press, New York 1995.

[12] S. K. Liaw, L. Dou and A. Xu, "Fiber-bragg-grating-based dispersion-compensated and gain-flattened raman fiber amplifier," Optics Express, vol. 15, pp. 12356–12361, 2007.

[13] L. G-Nielsen, M. Wandel, P. Kristensen, C. Jørgensen, L. Vilbrad Jørgensen, B. Edvold, B. Pálsdóttir, and D. Jakobsen, "Dispersion-Compensating Fibers," Journal of Lightwave Technology, vol. 23, pp. 3566–3579, 2005.

[14] E. M. Dianov, "Advances in Raman fibers," Journal of Lightwave Technology, vol. 20, pp. 1457–1462, 2002.

[15] L. Kazovsky, S. Benedetto, A. Willner, Optical fiber Communication Systems, 1st ed. Artech House Publishers, Norwood, 1996.

[16] L. Dou, M. Li, Z. Li, A. Xu, C.-Y. David Lan and S.-K. Liaw, "Improvement in characteristics of a distributed Raman fiber amplifier by using signal-pump double-pass scheme," Optical Engineering, vol. 45, No. 094201 (2006).

[17] G. P. Agrawal, *Nonlinear Fiber Optics*, 3rd ed. (Academic, New York, 2001).

[18] M. Tang, Y. D. Gong, and P. Shum, "Design of Double-Pass Dispersion-Compensated Raman Amplifiers for Improved Efficiency: Guidelines and Optimizations," Journal of Lightwave Technology, vol. 22, pp. 1899–1908, 2004.

[19] C. J. Anderson, and J. A. Lyle, "Technique for evaluating system performance using Q in numerical simulations exhibiting intersymbol interference," Electronics Letters, vol. 30, pp. 71–72, 1994.

7

High-Sensitivity Pressure, DP, and Random Rotational Angle Fiber Sensors

Wen-Fung Liu and Hao-Jan Sheng

Department of Electrical Engineering, Feng Chia University,
100 Wenhwa Road, Seatwen, Taichung, Taiwan 407,
Republic of China

Abstract

This chapter is devoted to three different fiber sensors based on the FBG for different sensing applications. The FBG is a kind of excellent optical passive component for a wide range of sensing applications by means of special structure designs and packages. This is due to the small size, immunity to electromagnetic interference, and in-line characteristics, etc. Here, we show several sensing applications by using FGBs, including a lateral pressure sensor, by combining the structure packaging design with the pressure sensitivity to be 10900 times higher than that of using a bare FBG, high-sensitivity temperature-independent DP sensor the sensitivity of 5.27×10^{-1} Mpa^{-1}, and a random rotational angle sensor using two fiber gratings with high accuracy for detecting the rotary position with full around 360° in any direction of a rotor, etc.

Keywords: FBG, sensing, pressure sensor, rotational angle sensor.

7.1 Fiber Grating Characterization and Fabrication

Fiber grating is one of the most optically important passive components with a broadrange of applications such as fiber sensors and communication systems. All-in-fiber configuration, small size, low insertion loss, and potentially low

Green Photonics and Smart Photonics, 115–140.

cost are the advantages of using fiber gratings over conventional technologies. The optical characteristics of a fiber grating can be particularly designed for different applications and requirements according to the design change both of grating period and grating index modulation. By adjusting the two parameters, specified spectral characteristics such as bandwidths and reflectivity can be obtained. For the fabrication of fiber gratings, the phase mask writing technique by a UV laser is the most popular method, although there are several other techniques to be demonstrated [1–7]. This method greatly simplified the fabrication process and yielded gratings with a nice repeatability and high performance [8]. The coupled-mode theory can be used for analyzing and can calculate the reflection and transmission spectra of gratings. The coupling between wave-guiding modes can be achieved with the effective refractive index perturbation in each wave-guiding mode when the phase-matching condition is satisfied. The phase-matching condition of a fiber grating means that the difference between the effective propagation constants of two distinct modes is equal to $2\pi/\Lambda$. This condition is also called the Bragg condition. For copropagating modes coupling through a uniform fiber grating, the phase-matching condition is given as $\beta_1 = \beta_2 = 2\pi/\Lambda$, where Λ is the grating period, and where $\beta_1 = 2n_{\text{eff1}}/\lambda$ and $\beta_2 = 2n_{\text{eff2}}/\lambda$ are the propagation constants of two wave-guide modes in the same propagation direction. This equation can be derived in terms of wavelength λ as $\lambda = (n_{\text{eff1}} + n_{\text{eff2}})\Lambda$. Thus, the coupling wavelength between the two counter-propagating LP_{01} core-modes is $\lambda_{\text{B}} = 2n_{\text{eff}}\Lambda$, where λ_{B} is called the Bragg wavelength. For example, in order to obtain Bragg reflection in the 1550 nm wavelength, the grating period should be about 535 nm for a typical single-mode fiber. For the coupling between the LP_{01} core-mode and a counter-propagating $\text{HE}_{1\mu}$ cladding-mode, the coupling wavelength is $\lambda_{\text{c}} = (n_{\text{eff01}} + n_{\text{eff1}})\Lambda$. For the coupling between the LP_{01} core-mode and copropagating $\text{HE}_{1\mu}$ cladding-mode, the interacting wavelength is $\lambda_{\text{r}} = (n_{\text{eff01}} - n_{\text{eff1}\mu})\Lambda$. Reflection and transmission spectra of fiber gratings can be theoretically derived from the coupled-mode theory. The grating maximum reflectivity is $R_{\text{max}} = \tanh^2(\kappa L)$, where κ and L are the grating coupling coefficient and the grating length, respectively. The reflectivity is proportional to the induced refractive index variations. Similarly, as the length of the grating increases, so does the resultant reflectivity. A calculated reflection spectrum as a function of the wavelength detuning can be obtained, in which the side lobes of the resonance are due to multiple reflections to and from opposite ends of the grating region.

7.2 The Lateral Pressure Sensors based on Fiber Gratings

A number of configurations of FBG pressure sensors have been experimentally demonstrated [9, 10]. Practically, sensitivity is an important specification for pressure sensors. Xu et al. [11] showed that a pressure sensitivity of the fractional change in the Bragg wavelength was -2.02×10^{-6} MPa^{-1}. In subsequent experiments [12], the pressure sensitivity was increased to -2.12×10^{-5} MPa^{-1} by using a glass-bubble housing for the FBG. Liu et al. later improved the pressure sensitivity to -6.28×10^{-5} MPa^{-1} by coating the FBG with a polymer [13]. Recently, Zhang et al. [14] further increased pressure sensitivity to as high as -3.41×10^{3} MPa^{-1} by using a shielded polymer-coated FBG. All of the above pressure sensors are operated by the axial compression strain of a FBG to cause the grating Bragg wavelength to be shifted toward the shorter wavelength side and then to obtain the negative pressure sensitivity. We proposed a new design mechanism for improving the pressure sensitivity of a FBG sensor to up to 2.2×10^{-2} MPa^{-1}. The positive sensitivity of this sensor is obtained and is based on the applied pressure to be transferred to the axial extended-strain on the FBG to cause the grating Bragg wavelength to be shifted toward the longer wavelength side.

7.2.1 Basic Operation Principle

The relation between the shift of the Bragg wavelength of FBG ($\Delta\lambda_B/\lambda_B$) and the axial strain ε applied to a fiber grating is [9]

$$\Delta\lambda_B/\lambda_B = (1 - p_e)\varepsilon, \tag{7.1}$$

where $P_e = n_{\text{eff}}^2 \left[P_{12} - \nu \left(P_{11} + P_{12} \right) \right]/2$ is the effective photo-elastic coefficient of the glass fiber with Poisson ratio ν, P_{11} and P_{12} denote the photo-elastic coefficients, and n_{eff} represents the effective refractive index of the guide mode, for example, a typical fused-silica fiber, $\nu = 0.16$, $n_{\text{eff}} = 1.46$, $P_{11} = 0.12$, $P_{12} = 0.27$, and thus, $P_e = 0.22$.

For a FBG coated with a thick polymer, and when both Young's modulus are close to each other, the axial strain along the FBG due to an applied pressure P is given by [15]

$$\varepsilon = -P(1 - 2\nu)/E, \tag{7.2}$$

where ν and E are the Poisson ratio and Young's modulus of the polymer, respectively.

For our packaged design, a FBG is encapsulated in a polymer-half-filled metal cylinder, the end of which is bound to the center of a round plate attached to the polymer surface. The other end of the FBG through the hole is glued at the left end of the cylinder, as shown in Figure 7.1.

The Young's modulus of the polymer is four orders lower than that of the FBG. This cylinder filled in a polymer has two openings on the opposite side of the wall, which can be pressurized mainly along one radial direction, and responds to an axial force acting on the round plate, creating an axial extended-strain on the FBG. The axial strain ε can be derived as

$$\varepsilon = \frac{\nu P A}{a E_{\text{FBG}} + \frac{L_{\text{FBG}}}{L_{\text{P}}}(A - a)E_{\text{polymer}}}, \tag{7.3}$$

where A denotes the round plate area, a represents the cross-section area of the FBG, P is the pressure acting on the polymer, ν denotes the polymer Poisson's ratio, L_{FBG} represents the FBG length, L_{P} is the axial length of the polymer, and E_{FBG} and E_{polymer} denote the Young's modulus of the FBG and the polymer, respectively. Hence, when pressure P is applied to this sensor, Equation (7.1) can be obtained as

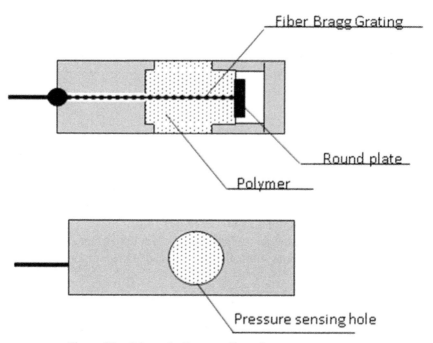

Figure 7.1   Schematic diagram of lateral pressure sensor.

$$\Delta\lambda_B/\lambda_B = (1 - P_e)\frac{\nu A P}{a E_{FBG} + \frac{L_{FBG}}{L_P}(A - a)E_{polymer}} = k_P P, \quad (7.4)$$

where $k_P = (1 - P_e)\frac{\nu A}{a E_{FBG} + \frac{L_{FBG}}{L_P}(A-a)E_{polymer}}$ is the pressure sensitivity of
the sensor. In the experiments, the used polymer is a kind of silicon rubber with
a Poisson's ratio of 0.4, the round plate has an area (A) of $(5)^2 \times \pi$ mm^2, the
cross-section area of the FBG (a) is $(0.0625)^2 \times \pi$ mm^2, the Young's modulus
of the FBG (E_{FBG}) and silicon rubber ($E_{polymer}$) are 7×10^{10} N/m^{-2} and
1.8×10^6 N/m^{-2}, respectively, and the ratio of L_{FBG} and L_P is 2. A pressure
sensitivity of $k_P = 2.15 \times 10^{-2}$ MPa^{-1} is thus obtained.

7.2.2 Experimental Results

The configuration of the pressure sensor is shown in Figure 7.1 in which two
holes are located opposite to one another on the sidewall of the cylinder part
of an Al cage. The center of a round plate is fixed to the end of a FBG. The
FBG is placed in the center of the Al cage from the open side, with the round
plate positioned on top of the two sidewall holes in the cylinder area. The
polymer is then poured into the cylinder from the open side, and it solidifies
inside the cylinder with the round plate of FBG attached to its surface of the
polymer. The FBG is fixed at the opening of the small hole with an adhesive
pretension, and the cylinder opening is sealed. The cylinder area has an inner
diameter of 11 mm, an inner length of 20 mm, a wall of 2 mm in thickness,
and two sidewall holes, each with diameters of 11 mm. The axial depth of the
small hole of the Al cage is 12 mm. Moreover, the round plate is Al and has
a diameter of 10 mm and a thickness of 0.5 mm. The FBG has a length of
20 mm. The polymer used is a type of silicon rubber. The sensor is put in a
temperature-controlled pressure chamber. The shift of the Bragg wavelength
corresponding to the variation in applied pressure is monitored with an optical
spectrum analyzer.

The optical spectra of pressure measurements in the range from 0 to
0.2 MPa is shown in Figure 7.2 where the unequal peak power levels result
from an ASE light source. The measured Bragg wavelength is a function
of the applied pressure shown in Figure 7.3, which also displays very good
linearity between the Bragg wavelength and the pressure. The shift of the
Bragg wavelength in response to the applied pressure is 33.876 nm/MPa,
corresponding to a pressure sensitivity of 2.2×10^{-2} MPa^{-1}. This pressure

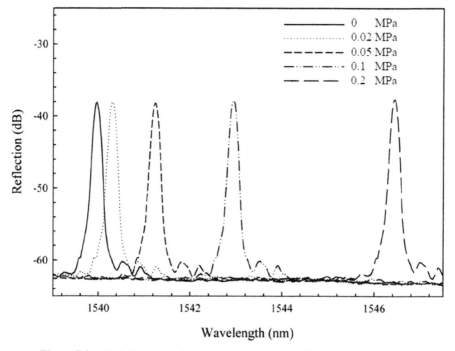

Figure 7.2   Optical spectra of pressure sensor measured from 0.02 to 0.2 MPa.

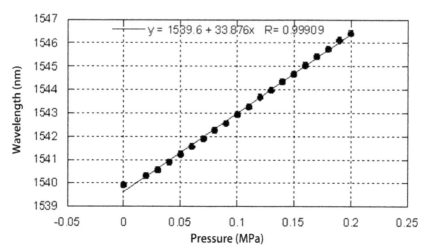

Figure 7.3   Measured Bragg wavelength as a function of the applied pressure.

sensitivity value is approximately 10,900 times larger than that measured with a bare FBG.

The cross-sensitivity to temperature for the designed sensor is approximately $7.624 \times 10^{-6}/°C$, which is obtained by the wavelength shift of 0.587 nm measured in the temperature range from 20 to 70°C. Because of the pressure sensitivity significantly exceeding the temperature sensitivity, thermal compensation can be neglected in the temperature variation.

The measured pressure sensitivity 2.2×10^{-2} MPa^{-1} closely approximates the theoretical value 2.15×10^{-2} MPa^{-1} calculated from Equation (7.4). In the ideal situation, the pressure applied to the sensor is based on single coordinate axial direction only. The error between the calculated and measured sensitivity is thus attributed to the pressure sensing holes being elliptic, and the applied pressure not coming from one coordinate axial direction only, causing the actual strain experienced by the silicon on the FBG to exceed the ideal. Therefore, this proposed pressure sensor based on a FBG is demonstrated with a sensitivity of 2.2×10^{-2} MPa^{-1}, which is approximately 10,900 times higher than that for a bare FBG. Positive sensitivity means that the applied pressure is transferred to the axial stretched-strain on the FBG. This sensor should have a wide range of applications, including in measuring medium pressure, liquid level, and underwater depth.

7.3 The DP Sensors based on Fiber Gratings

Differential pressure operating mechanisms for high pressure and liquid-flowing rate detection can be achieved by using FBG pressure sensors [16, 17]. For air pressure and temperature control of cabin in an airplane, submarine, and laboratory or for the production room in a pharmaceutical and electronic factory, the measurement sensitivity of DP and temperature sensing device is an important factor for providing a comfortable and safe working place. In this study, a new designed DP fiber sensor based on FBGs has been demonstrated, in which two identical FBGs are packaged in a temperature-compensated metal structure. This DP sensor based on FBGs is experimentally demonstrated to have the maximum sensitivity of 821.87 nm/MPa. Therefore, by means of this simple packaging construction, high-sensitivity temperature-independent DP sensor can be achieved with a nice linearity for extending the applications in the measurement of liquid level, liquid density, or specific gravity detection.

7.3.1 Basic Principle

The sensor has a metal cage composed of two metal cylinders, in which a diaphragm consists of a silicone rubber sandwiched between two metal plates located in the middle of sensor structure. Thus, this sensor includes two symmetrical pressure cavities separated by the diaphragm, as shown in Figure 7.4.

There is a pressure inlet hole on each pressure cavity. Two identical FBGs with the grating length of 1 cm and the separation of 0.8 cm are fabricated in a SMF-28 fiber by phase-mask writing techniques. They are glued axially through the center of diaphragm at each pressure cavity with sufficient pre-strain (ε_s) to accommodate the wavelength shift induced by DP. Surrounding temperature variation would cause a thermo-expansion of metal drum which induces the strain of $\varepsilon_T = M \times \Delta T$ that is identical for both FBGs, where M is the thermal expansion coefficient of metal and T is the temperature. When a liquid or an air with different pressure comes into the sensor through two pressure inlet holes respectively, a net pressure will act on the surface of round plate and diaphragm. This net pressure will induce an axial tension force to the FBG in the higher pressure cavity and an axial compress strain in the lower pressure cavity. A strain of FBG induced in a pressurized cavity with a diaphragm-hardcore is given by [18]

$$\varepsilon_P = C \times P, \tag{7.5}$$

where $C = \dfrac{\frac{R^4}{64D}\left[1-\left(\frac{r}{R}\right)^4+4\left(\frac{r}{R}\right)2\ln\frac{r}{R}\right]}{L+\frac{a_f E_f R^2}{16\pi D}\left[1-\left(\frac{r}{R}\right)^2\frac{1-\left(\frac{r}{R}\right)^2+4\ln^2\left(\frac{r}{R}\right)}{1-\left(\frac{r}{R}\right)^2}\right]}$, $D = \dfrac{Et^3}{12(1-\nu^2)}$, P is the

pressure in the sensor, R is the diaphragm radius, t is the diaphragm thickness, r is the round plate radius, a_f is the fiber cross-section, E_f is the Young's

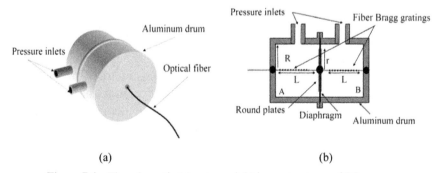

(a)   (b)

Figure 7.4   The schematic (a) outer and (b) inner structures of DP sensor.

modulus of fiber, and E and v are the Young's modulus and the Poisson's ratio of the diaphragm, respectively.

In our proposed sensor, the total strain of FBG in the higher pressure cavity can be expressed as

$$\varepsilon_{f,H} = \varepsilon_{p,H} + \varepsilon_T + \varepsilon_s, \tag{7.6}$$

where $\varepsilon_{p,H}$ is the strain induced only by the pressure in the higher pressure cavity. The total strain of FBG in the lower pressure cavity is given as

$$\varepsilon_{f,L} = -\varepsilon_{p,L} + \varepsilon_T + \varepsilon_s, \tag{7.7}$$

where $\varepsilon_{p,L}$ is the strain induced only by the pressure in the lower pressure cavity. The Bragg wavelength shift caused simultaneously by strain and temperature variation can be expressed as [19]

$$\Delta\lambda_B = \left((1 - P_e)\varepsilon + \left[\alpha + \frac{\frac{dn}{dT}}{n} \right] \Delta T \right) \lambda_B, \tag{7.8}$$

where P_e is the effective photo-elastic constant of fiber. The center wavelength separation between two FBGs caused only by DP (ΔP) is expressed as

$$\frac{\Delta\lambda_{B,H} - \Delta\lambda_{B,L}}{\lambda_B} = (1 - P_e)2C \times \Delta P, \tag{7.9}$$

where $\Delta\lambda_{B,H}$ and $\Delta\lambda_{B,L}$ are the reflected central wavelength of the gratings in the higher and lower pressure cavities, respectively. The shift of the center-wavelength average-value of two FBGs caused by the temperature variation of ΔT is obtained as

$$\frac{\Delta\lambda_{B,H} + \Delta\lambda_{B,L}}{2\lambda_B} = (1 - P_e)\varepsilon_s + \left((1 - P_e)M + \alpha + \frac{\frac{dn}{dT}}{n} \right) \Delta T. \tag{7.10}$$

From Equation (7.9), the DP to be measured is proportional to the separation of the central wavelengths of two FBGs. From Equation (7.10), the temperature can be determined by the average value of both reflected central wavelength shifts in the sensor. Therefore, this sensor has the capability to measure DP and temperature simultaneously.

7.3.2 Experimental Results

The physical configuration of the sensor is made of Al. The original reflected central wavelengths of both FBGs in the sensor are equally located in

1558.74 nm at 15°C with a little bit different reflectivity. Thus, the FBG in each cavity can be identified in the sensor. They are prestrained for the reflected central wavelength to be shifted to 1561.76 nm and the reflective peaks are kept in overlap condition during the fabricating process. Hence, there is only one reflective peak shown on optical spectrum analyzer before the sensor is pressurized. All structural parameters of the sensor are shown in Table 7.1. The experimental setup is shown in Figure 7.5.

The shift of the reflected central wavelength corresponding to the variation of DP is monitored by using an optical spectrum analyzer. The pressure air is supplied by a piston type pump which can easily provide a positive and

Table 7.1　Parameters of sensor structure

Parameters	Value	Parameters	Value
a_f	0.0123 mm^2	E_f	70 GPa
L	20 mm	R	12.5 mm
M	$23.1 \times 10^{-6}/°C$	E	1.8×10^6 N/m^{-2}
t	1 mm	r	10 mm
v	0.4	λ_B	1558.74 nm

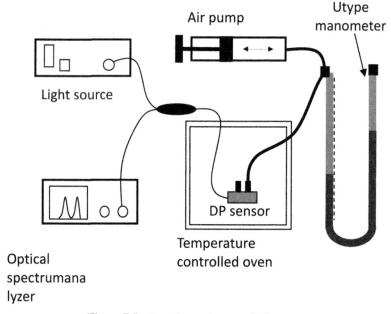

Figure 7.5　Experimental setup of DP sensor.

negative pressure in the sensing device. A U-type manometer is connected to the air pressure loop for pressure measurement. The air with different pressure is applied to one pressure inlet of the sensor, which then is put in a temperature-controlled oven for monitoring the reflected central wavelengths shift with temperature variation. Hence, different pressure can be applied to this sensor at any temperature for confirming the pressure and temperature measurement simultaneously.

For the temperature measurement at no DP, the wavelength shift of the sensor versus temperature variation from 15 to 60°C is shown in Figures 7.6 and 7.7. There is only one reflective peak in the reflected central wavelengths shift profile when ambient temperature is varied.

The responsivity of reflected central wavelengths shift versus temperature is around 0.04 nm/°C with a nice linearity and without obvious deviation. Thus, a symmetrical packaged structure in the higher and lower pressure cavity of the sensor is experimentally confirmed. In the sensor, the cavity with higher

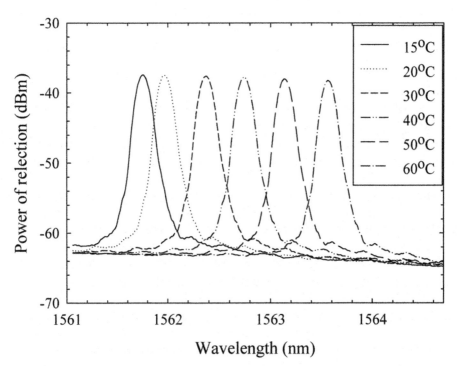

Figure 7.6 The spectral profile of wavelength shift versus temperature variation.

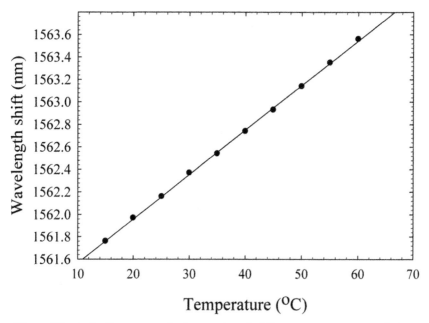

Figure 7.7   A nice linear curve for the wavelength shift versus temperature variation.

reflected peak FBG is denoted as A-cavity and the other is B-cavity. During DP measuring experiments, A-cavity is pressurized by a piston pump at different surrounding temperature and the pressure inlet of B-cavity is vented. The pressure applied to A-cavity is in the range of +50 to –50 cmH$_2$O with a decrement of 10 cmH$_2$O, and the surrounding temperature is 15 to 60°C with an increment of 5°C. When DP is changed at fixed surrounding temperature, the reflected central wavelengths of the two FBGs in respective cavities shifted in opposite direction with an equal shifting amount. The profiles of reflected central wavelengths shift with DP variation at temperature 25 and 45°C are shown from Figures 7.8–7.11. When the surrounding temperature is changed at a fixed DP stage, reflected average-central wavelength shifts linearly to temperature variation. The wavelength of two reflected center peaks at DPs (P_A – P_B) of respectively 20 and 50 cmH$_2$O with surrounding temperature variation of 15–60°C in an increment of 5°C are shown in Figure 7.12. The separation between reflected central wavelengths of both FBGs is only proportional to Dp without any interference of temperature perturbation. A measuring responsivity of 0.806 nm per 10 cmH$_2$O is obtained in the experiment as shown in the curve slope of Figure 7.13.

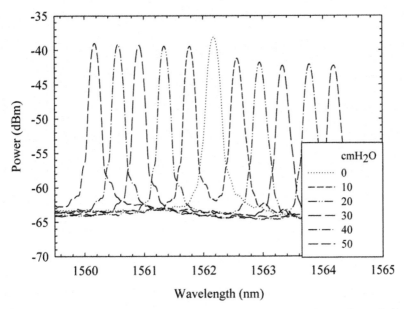

Figure 7.8 The profile of the FBGs wavelength shift with DP (cmH$_2$O) variation at temperature 25°C, $P_A < P_B$.

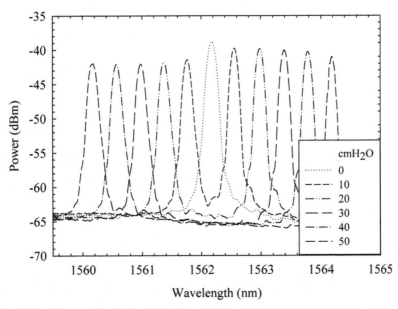

Figure 7.9 The profile of the FBGs wavelength shift with DP (cmH$_2$O) variation at temperature 25, $P_A > P_B$.

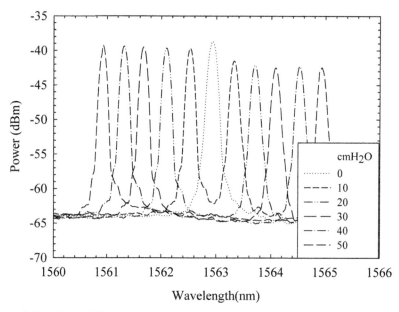

Figure 7.10 The profile of the FBGs wavelength shift with DP (cmH_2O) variation at temperature 45°C, $P_A < P_B$.

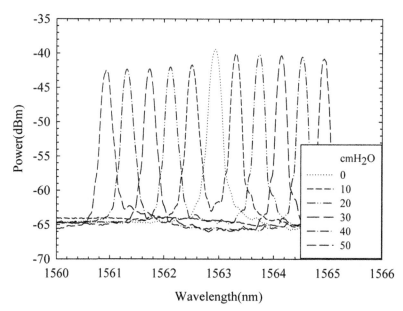

Figure 7.11 The profile of the FBGs wavelength shift with DP (cmH_2O) variation at temperature 45°C, $P_A > P_B$.

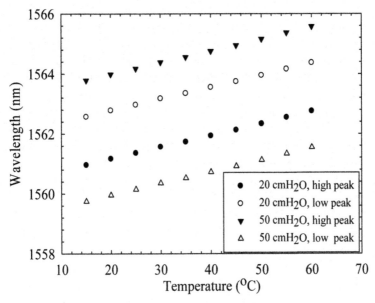

Figure 7.12 Position of two wavelength peaks at pressure $P_A - P_B = 20$ and 50 cmH$_2$O with temperature variation of 15–60°C in increment of 5°C.

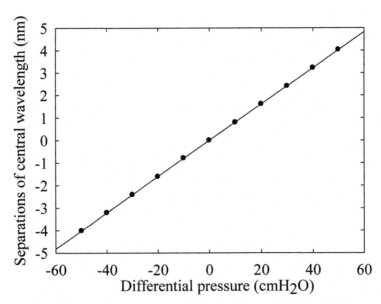

Figure 7.13 The separation between reflected central wavelengths of FBGs versus DP $(P_A - P_B)$ at any temperature with a responsivity of 0.806 nm/10 cmH$_2$O.

Thus, the DP sensitivity of 821.87 nm/MPa is equivalent to the DP sensitivity factor of 5.27×10^{-1}/MPa (normalized value with 1558.74 nm). According to Equation (7.9), the theoretical responsivity is 0.915 nm per 10 cmH$_2$O. The discrepancy between the experimental result and the theoretical value is attributed to the dimension error of each element in the sensor.

In the proposed sensor, the DP measuring range is determined by the maximum allowable stretched strain of FBG located in the high-pressure cavity and the prestrain of FBG located in the low-pressure cavity at certain temperature. The maximum allowable stretched strain of an FBG fabricated in SMF-28 is around 8000 micro-strain. In the temperature range of our experiment, the maximum DP measuring range can be −11.681 to +11.681 kPa (−119 to +119 cmH$_2$O) at 59.2°C, and the minimum measuring range is −7.349 to + 7.349 kPa (−74.9 to + 74.9 cmH$_2$O) at 15°C. For the FBGs with an appropriate prestrain, the maximum measuring range of DP can be obtained at a certain required ambient or working temperature.

7.4 The Rotational Angle Sensors based on Fiber Gratings

For the subject of rotary position sensors in mechanical systems, Montanini et al. [20] have shown that using two FBGs and the tangential strain can be measured to sense the rotary position of a rotor attached to the central end of an approximated Archimedes spiral. Nishiyama et al. proposed that macro-bending of a hetero-core fiber driven by a preshaped disk on the end of a rotor. The rotary position can be determined by the light-intensity variation of the hetero-core fiber output [21]. Baptista et al. [22] had proposed a sensing head composed of a long-period fiber grating (LPG) whose peak-power variation is a function of angular rotation. The methods above mentioned cannot have the function of detecting a rotor to be rotated at the same direction infinitely. For improving this disadvantage to be practically applied in meteorological anemoscope, ship's rudder indicator, airfoil control, and other relevant servomechanisms, the rotary position sensors should have capability to detect the rotating angle of a rotor which is rotated infinitively at arbitrary direction. Therefore in this paper a random rotational angle sensing structure based on a pair of cantilever-driven FBGs is proposed with the function to detect the rotating angle of a rotor to be rotated at arbitrary direction and rotated at the same direction infinitely. Owing to its special design structure, the physical size of the proposed sensor can be reduced on a requirement and beneficial for fitting in small rotary apparatus functioning as a rotating-positional sensor.

7.4.1 Basic Operating Principle

The sensor is composed of a grating sensing head shielded by an Al cylinder, as shown in Figure 7.14a. From this figure, we can see that one Al shaft with the disc driven by input rotating activation is located at the right-hand side. One end of a metal-rod beam with 1.2-mm diameter is fixed axially at the center of non-shaft side in the Al cylinder, and a 2-mm diameter magnet is mounted on the other end close to the Al disc. The 20-mm length shaft which is elongated inside the cylinder ends with a 7-mm diameter Al disc mounted with two 2-mm diameter magnets. These two magnets not only repel each other but also provide the resultant magnetic repulsion to cause the thin metal rod to be bent as a cantilever, as shown in Figure 7.14b.

The sensing principle is based on the wavelength shift of two identical FBGs glued on the metal-rod surface. These two FBGs are bound axially on the metal rod apart from each other by a quarter of circumference, as shown in Figure 7.15.

The relation between the shift of the Bragg wavelength of FBG ($\Delta\lambda_B/\lambda_B$) and the axial strain ε applied to a fiber grating is [9]:

$$\frac{\Delta\lambda_B}{\lambda_B} = (1 - P_e)\,\varepsilon, \qquad (7.11)$$

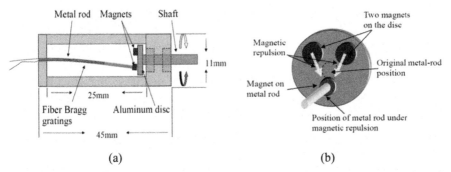

(a)   (b)

Figure 7.14   (a) Rotational schematic diagram of the sensor, (b) magnetic repulsion activated to magnet on metal rod.

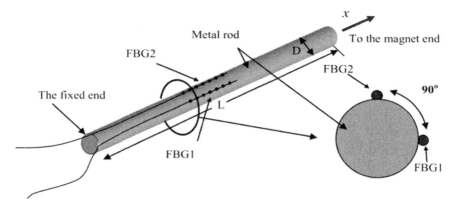

Figure 7.15   Two FBGs are bound axially on the metal rod apart from each other by 90°.

where P_e is the effective photo-elastic coefficient of the glass fiber, for a typical fused-silica fiber, $P_e = 0.22$. For a FBG located at the middle of a simple micro-cantilever and to neglect the gravity both of the mass of the cantilever and the used magnet, the average axial strain ($\varepsilon_{\text{beam}}$) in terms of the maximum deflection δ_{max} at the free end can be derived as

$$\varepsilon_{\text{beam}} = \frac{3D\delta_{\text{max}}}{4L^2}, \tag{7.12}$$

where L is the beam length and D is the beam diameter.

When the shaft in the proposed sensor is rotated and the two magnets go around the magnet at the end of the metalrod, the direction of the magnetic repulsion would change and will be dependent on the rotating position, but the amount of magnetic repulsive force on the free end of the thin metalrod is same all the time. Because the metalrod is symmetrical around the central axis, the bending direction of the thin metalrod changes following the rotating position variations, yet the bending curvature remains the same. Hence, the maximum axial strain ($\varepsilon_{\text{beam}}$) on the bent metalrod is always the same and opposite to the bending direction. The axial strain of a FBG glued on the metalrod is varied with bending directions, as shown in Figure 7.16.

The axial strain on FBG can be concluded as

$$\varepsilon_{\text{FBG}} = \sin\theta \times \varepsilon_{\text{beam}}, \tag{7.13}$$

Where θ is the rotating angle of sensor shaft.

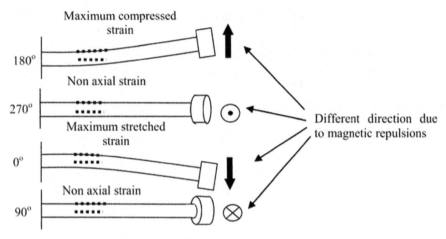

Figure 7.16   The axial strain of a FBG on the metalrod is varied with bending directions.

To be a rotary sensor by using this structure based on two FBGs (FBG1 and FBG2), the rotating angle can be derived as

$$\theta = \cos^{-1}\left[\left(\frac{\Delta\lambda_2}{\lambda_{max}} - \frac{\Delta\lambda_1}{\lambda_{max}} \times \cos\varphi\right)\frac{1}{\sin\varphi}\right] \times \frac{\Delta\lambda_1}{|\Delta\lambda_1|}$$
$$+ \left(1 - \frac{\Delta\lambda_1}{|\Delta\lambda_1|}\right) \times 180^\circ, \tag{7.14}$$

where $\Delta\lambda_1$ and $\Delta\lambda_2$ is the central wavelength shift of FBG1 and FBG2, respectively, and λ_{max} is the average maximum central wavelength of two FBGs. φ is the actual phase difference between FBG1 and FBG2 obtained by the experimental results.

7.4.2 Experimental Results

Our proposed rotary position sensor is composed of two FBGs to be glued on a 1.2-mm diameter spring steel rod at one end with apart from each other ideally by 90°. The two identical FBGs with the grating length of 10 mm are fabricated in SMF-28 fiber with the same central wavelength of 1553.20 nm to minimize the chirp effect [23]. Two strong magnets of 2 mm diameter are glued at the same side on an Al disc attached to the shaft. All the magnets in the structure belong to Neodymium Iron Boron. Then, the spring steel rod assembly is fixed to close (side) end center in the Al cylinder. The rotary shaft with open end cover and two steel ball bearings is mounted on the open end of the Al cylinder. The diameter and the length of the Al cylinder are 11 and 45 mm, respectively.

The experimental setup for measuring the angular position is shown in Figure 7.17. While the shaft is driven by a rotor to be rotated, the central wavelengths of two FBGs (FBG1 and FBG2) bound on thin metal rod surface should be theoretically shifted to show the sinusoidal function variation by a phase difference of 90°.

To verify the capability of detecting the random rotating position for the proposed sensor, the sensor shaft is rotated in the way of alternating the clockwise and anticlockwise rotation by an increment of 15° each time and the wavelength shifts both of FBG1 and FBG2 are measured. The initial point (0° position) of the sensor shaft is the position with the maximum grating wavelength shift of FBG2 but no wavelength shift of FBG1. First to turn clockwise the shaft to 15°, the wavelength shifts both of FBG1 and FBG2 are obtained. Then, the shaft is rotated anticlockwise to 30° for recording the two grating wavelength shifts. And then to rotate clockwise the shaft to 45° for the measurement of the two grating wavelength shifts, in the same way we can obtain all the data of the rotating positions of 360°. The reflective spectra

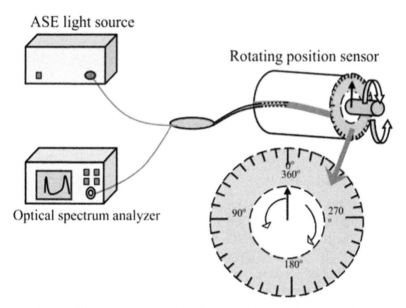

Figure 7.17   Experimental setup for monitoring the rotating position.

both of FBG1 and FBG2 for the increment of 45° are, respectively, shown in Figure 7.18. We can see that the reflective spectra both of FBG1 and FBG2 are shown from Figures 7.18a–h for the rotating angles from 0°–315° with the increment of 45°.

The grating wavelength shifts both of FBG1 and FBG2 versus to the sensor shaft rotating positions at room temperature are shown in Figure 7.19. We can see that the wavelength shifts of the two FBGs are sinusoidal variation and dependent on the shaft rotating position as the theoretical prediction. However, the metal rod is so small in diameter that it is difficult to locate the two FBGs exactly a quarter of circumference apart on the surface of rod. The angle interval between two maximum wavelength shifts of FBG1 and FBG2 is used to make sure the two sinusoids wave function to have a phase difference of 90°.

The absolute value of the average maximum central wavelength of the two FBGs is 1.7 nm. The experimental results of the 107° separation between

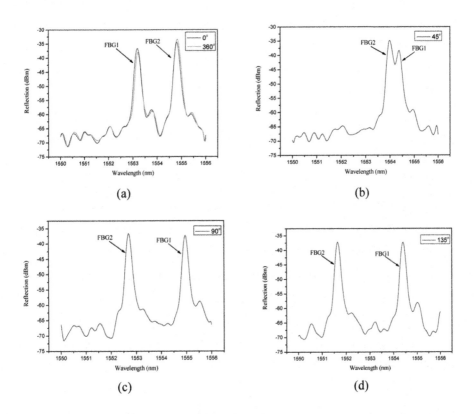

(a)   (b)

(c)   (d)

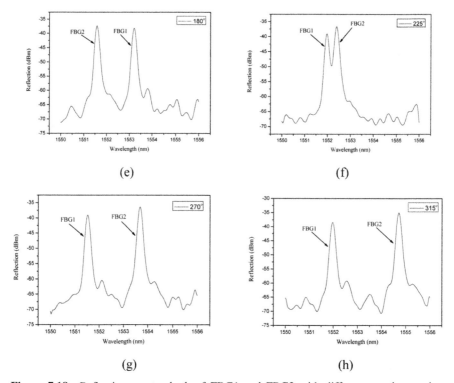

(e)

(f)

(g)

(h)

Figure 7.18   Reflection spectra both of FBG1 and FBG2 with different rotating angles, (a) 0° 360°, (b) 45°, (c) 90°, (d) 135°, (e) 180°, (f) 225°, (g) 270°, and (h) 315°.

two maximum wavelength shifts of FBG1 and FBG2 show the fact that the actual positions of these two FBGs are bound apart by more than a quarter of circumference. This is because the metal rod is so thin that it is difficult to glue the two FBGs just apart from each other by 90° precisely. Substituting all the wavelength shifts into Equation (7.14), the relationship between the demodulated shaft angle and real angle is shown in Figure 7.20a and a deviation of each demodulated angle is shown in Figure 7.20b.

A deviation of only 1.1° with the temperature variation from 19 to 45°C has been experimentally demonstrated. This small deviation is attributed to the wavelength shift of two FBGs induced by the temperature variation compensated by axial thermal expansion of spring steel rod that just makes its bending curvature smaller.

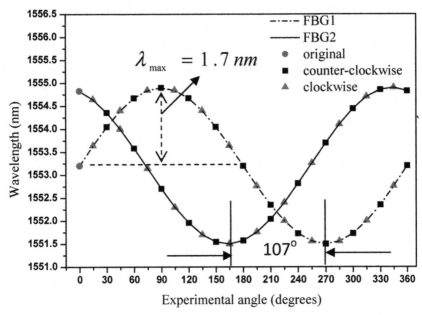

Figure 7.19 The grating wavelength shifts both of FBG1 and FBG2 versus the shaft rotating positions.

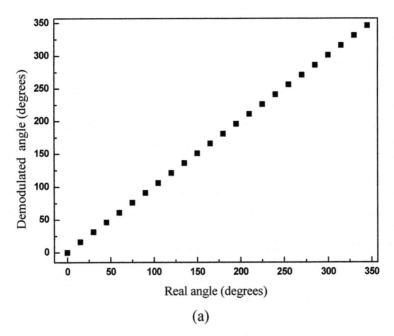

(a)

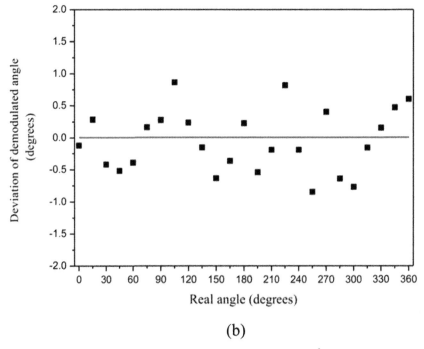

(b)

Figure 7.20   Experimental results of rotary angle detecting at 20°C: (a) the relationship between real angle and demodulated angle, (b) deviation of demodulation angle.

7.5 Conclusion

In this chapter, there are three different types of fiber sensors to be investigated. The first one is a lateral pressure sensor based on a FBG to be demonstrated with the sensitivity of 2.2×10^{-2}/MPa approximately 10900 times higher than that for a bare FBG. This sensor can be applied in measuring medium pressure, liquid level, and underwater depth. The second type is a DP sensor by means of novel packaged-structure design with the sensitivity of 5.27×10^{-1} Mpa^{-1}. This device can be used for measuring DP and temperature simultaneously specially in liquid level, liquid density, or specific gravity detection. The third type is a rotational position sensor based on two fiber gratings with characteristics of linearity, stability, negligible hysteresis phenomenon. With its tiny size, light weight, and arbitrary random rotational angle position detecting ability, this sensor has a potential application in small rotary devices equipped within a limited space and is expected to be used for a rotor position detecting on an aero/space platform.

References

[1] K. O. Hill, Y. Fujii, D. D. Johnson, and B. S. Kawasaki, "Photosensitivity in optical fiber waveguides: Application to reflection filter fabrication," Appl. Phys. Lett., 32, pp. 647–649, 1978.

[2] G. Meltz, W. W. Morey, and W. H. Glenn, "Formation of Bragg gratings in optical fibers by a transverse holographic method," Opt. Lett., 14, pp. 823–825, 1989.

[3] C. G. Askins, T. -E. Rasi, G. M. Williams, M. A. Putnam, M. Bashkansky, and E. J. Friebele, "Fiber Bragg reflectors prepared by a single excimer pulse," Opt. Lett., 17, pp. 833–835, 1992.

[4] B. J. Eggleton, P. A. Krug, L. Poladian, K. A. Ahmed, and H. F. Lin, "Experimental demonstration of compression of dispersed optical pulses by reflection form self-chirped optical fiber Bragg grating," Opt. Lett., 19, pp. 877–879, 1994.

[5] Q. Zhang, D. A. Brown, L. Reinhart, and T. F. Morse, "Simple prism-based scheme for fabricating Bragg gratings in optical fibers," Opt. Lett., 19, pp. 2030–2032, 1994.

[6] X. Wang, Y. Liao and K. Zha, "Theoretical and experimental study on the fabrication of double fiber grating," Opt. Fiber Tech., 3, pp. 189, 1997.

[7] H. Patrick. and S. L. Gilbert, "Growth of Bragg gratings produced by continuous-wave ultraviolet light in optical fiber," Opt. Lett., 18, pp. 1484–1486, 1993.

[8] K. O. Hill, B. Malo, F. Bilodeau, D. C. Johnson, and J. Albert, "Bragg gratings fabricated in monomode photosensitive optical fiber by UV exposure through a phase mask," Appl. Phys. Lett., 62, pp. 1035–1037, 1993.

[9] W. W. Morey, G. Meltz, and W. H. Glenn, "Fiber optic Bragg grating sensors," in Proc. SPIE, Fiber Optics and Laser Sensors VII, Vol. 1169, pp. 98–107, 1989.

[10] A. D. Kersey, M. A. Davis, H. J. Patrick, M. LeBlanc, K. P. Koo, C. G. Askins, M. A. Putnam, and E. J. Friebele, "Fiber grating sensors," J. Lightwave Technol. Vol.15, pp. 1442–1463, 1997.

[11] M. G. Xu, L. Reekie, Y. T. Chow, and J. P. Dakin, "Optical in-fiber grating high pressure sensor," Electron, Lett., Vol. 29, pp. 398–399, 1993.

[12] M. G. Xu, H. Geiger, and J. P. Dakin, "Fiber grating pressure sensor with enhanced sensitivity using a glass-bubble housing," Electron, Lett., Vol. 32, pp. 128–129, 1996.

[13] Y. Liu, Z. Guo, Y. Zhang, K. S. Chiang, and X. Dong, "Simultaneous pressure and temperature measurement with polymer-coated fiber Bragg grating," Electron, Lett., Vol. 36, pp. 564–566, 2000.

[14] Y. Zhang, D. Feng, Z. Liu, Z. Guo, X. Dong, K. S. Chiang, and Beatrice C. B. Chu, "High-sensitivity pressure sensor using a shielded polymer-coated fiber grating," Photon. Technol. Lett., Vol. 13, pp. 618–619, 2001.

[15] G. B. Hocker, "Fiber-optical sensing of pressure and temperature," Appl. Opt., Vol. 18, pp. 1445–1448, 1979.

[16] Y. Zhao, C. Yub, Y. Liao, "Differential FBG sensor for temperature compensated high pressure (or displacement) measurement," J. Optics and Laser Technol., 36, pp. 39–42, 2004.

[17] J. Lim, Q. P. Yang, B. E. Jones and P. R. Jackson, "DP flow sensor using optical fiber Bragg grating" J. Sensors and Actuators, 92, pp. 102–108, 2001.

[18] W. T. Zhang, F. Li, Y. L. Liu, and L. H. Liu," Ultrathin FBG Pressure Sensor with Enhanced Responsivity" IEEE Photon. Tech. Lett., 19, pp. 1553–1555, 2007.

[19] A. D. Kersey, M. A. Davis, H. J. Patrick, M. LeBlanc, K. P. Koo, C. G. Askins, M. A. Putnam and E. J. Friebele "Fiber Grating Sensors" J. Lightwave Technol., 15, pp. 1442–1463, 1997.

[20] R. Montanini and S. Pirrotta, "A temperature-compensated rotational position sensor based on fibre Bragg gratings," Sens. Act. A, 132, pp. 533–540, 2006.

[21] M. Nishiyama and K. Watanabe, "Rotational positioning measurement for the absolute angle based on a hetero-core fiber optics sensor," Proc. SPIE, 7503, 2009.

[22] J. M. Baptista, S. F. Santos, G. Rego, O. Frazão and J. L. Santos, "Measurement of Angular Rotation using a Long Period Fiber Grating in a Self-Referenced Fiber Optic Intensity Sensor," in Proceedings of IEEE Conference on Laser and Electro-Optics Society, pp. 806–807, 2005.

[23] D. H. Kanga, S. O. Parkb, C. S. Hongb, and C. G. Kimb, "The signal characteristics of reflected spectra of fiber Bragg grating sensors with strain gradients and grating lengths," NDT and E Int., 38, 712–718, 2005.

8

Heterogeneous Integration of Group IV Semiconductors on Si by RMG Method for Implementing High-Speed Optoelectronic Devices

Chih-Kuo Tseng, Wei-Ting Chen, Ku-Hung Chen, Ching-Hsiang Chiu, Shih-Che Yen, Neil Na and Ming-Chang M. Lee

Institute of Photonics Technologies and Department of Electrical Engineering, National Ting-Hua University, Hsinchu, Republic of China

Abstract

Si-based high-speed optoelectronic devices are keystones for over 100 Gbps data communication potentially applied in data centers, high-performance cluster computing, and cloud computing servers. A unique process using the RMG method, in combination with the self-aligned microbonding technique, is applied to heterogeneously integrate monocrystalline Group IV semiconductor on Si photonic devices. This process doesn't require complex epitaxy process steps to deal with the lattice mismatch issue between different semiconductors, and the thermal budget is relatively small, which is most compatible with the standard CMOS process. Several waveguide-based high-speed Si/Ge/Sn photodetectors are presented, including Si/Ge heterojunction waveguide pin, Si/Ge butt-coupling waveguide photodetectors, and GeSn photodetectors.

Keywords: SOI, MEMS, SAMB.

8.1 Introduction

Owing to the increasing demand in transmitting broadband data through optical fibers for short-range interconnect or long-haul telecommunication, cost-effective and high-performance integrated high-speed optoelectronic

Green Photonics and Smart Photonics, 141–178.

devices become a key element in boosting market growth. Silicon or germanium-based photonics fabricated through CMOS-compatible process could help achieve this goal due to the relatively low material cost, advanced process technology, and high volume production. Several devices such as high-speed modulators, multiplexing and demultiplexing passive lightwave circuits, and Si or Ge photodetectors have been reported with excellent performance in recent years. For near-infrared photodetection, Si (bandgap ∼1.12 eV) is not a suitable material because it is transparent at the wavelength band for optical fiber communication. To solve the problem of inefficient light absorption, Ge (bandgap ∼0.66 eV) is often heterogeneously integrated on Si substrate for optical receiver applications.

In the aspect of material growth, the most common method used is a direct epitaxy of Ge on Si despite their 4% lattice mismatch that easily generates threading dislocations at the interface. Serious negative impacts on photodetector performance such as large leakage current and reduced photo-responsivity may occur, and therefore fine-tuning the epitaxial recipe to optimize the growth process is rather important. Typically, a two-step temperature-varying growth process is used and a high thermal budget is often required [1], which could be harmful for front-end CMOS process steps. In contrast to epitaxial growth, Liu et al. [2] first proposed and studied the RMG Ge that features low thermal budget and is fully compatible with CMOS fabrication process. They demonstrated a lateral growth of single-crystalline Ge strip on insulator by RTA at a temperature >937°C for a few seconds. High-quality Ge can be obtained at places away from the Si seed window, which traps and terminates the defects that arise the lattice mismatch at Ge–Si interface. Using this method, Assefa et al. [3–5] demonstrated a novel MSM waveguide photodetector on SOI substrate, but unfortunately their devices showed a rather large leakage dark current.

In this chapter, we first introduce how the RMG method can be used to grow single crystal germanium on Si substrate and how it can be applied for implementing Ge MSM photodetectors with surface passivated by amorphous silicon, in Section 8.1. In Section 8.2, we present a novel process called SAMB technique to obtain high-quality Ge/Si heterogeneous junctions, which is applied for high-speed waveguide-based Ge/Si photodiodes. Here, surface tension force is utilized to make Ge/Si heterojunctions by releasing as-grown RMG Ge beams on Si structures separated by SiO_2. This technique is indeed simple, is of low cost, of low thermal budget, and is compatible with a wafer-level process, compared to commercial epitaxial growth or wafer bonding.

Furthermore, we apply the SAMB process on butt-coupled Ge photodetectors (Section 8.3). Butt-coupling between silicon waveguides and Ge absorbers is with a better coupling efficiency than evanescent coupling, especially for thick waveguides. We also investigate RMG of GeSn alloy in Section 8.4 and demonstrate a GeSn photodetector showing a better long-wavelength absorption efficiency than pure Ge devices.

8.2 Germanium MSM Photodetectors

Among various photodetector structures [6–12], MSM photodetectors are perhaps the simplest one with device fabrication possibly best compatible with the standard CMOS process. In this section, we design and fabricate a normal-incidence, MSM photodetector using RMG Ge in contact with Al on Si substrate. However, for most Ge devices, a large leakage current associated with metal-Ge contacts (Fermi-level pinning) and Ge surfaces (broken/dangling bonds) is usually exhibited without a proper surface or interface treatment. Many approaches are proposed to overcome this problem, for example, the use of barrier height enhancement [13], asymmetric contacts [14], and surface passivation [15]. We show that when applying an a-Si surface passivation [16, 17], the leakage dark current is reduced to the order of μA. Such a low-cost, high-performance MSM photodetector fabricated with a low thermal budge can be a vital solution to the integration of Ge photodetectors with ICs such as TIAs and beyond in a CMOS fabrication line.

In Figure 8.1, we illustrate the process flow of making the devices. In Figure 8.1a, a 6-inch Si substrate is cleaned and a SiNx of 150 nm is deposited by LPCVD. A single-crystalline Si seed window is opened by wet etching SiNx. Then, a Si adhesion layer of 3 nm and an amorphous Ge (a-Ge) layer of 300 nm are deposited by e-gun evaporator. The Ge strips are patterned by dry etch with Cl_2 and HBr chemistry. A cap oxide is coated by PECVD, followed by a RTA at 950°C for 4 s. The melted Ge recrystallizes laterally starting from the Si seed window when the system cools down, and the misfit dislocation defects at Ge–Si interface are necked around the corner of the Si seed window. Next, the capped oxide is removed by diluted BOE solution where Ge strips are retained on the SiN layer, as shown in Figure 8.1b. Then, in order to suppress the leakage current arisen from the dangling bonds on Ge surfaces, a 30-nm thick a-Si layer is deposited for surface passivation. A thin oxide layer is used for ILD and then a via-hole is dry-etched for exposing the Ge surface. A lift-off process is used for Al metallization that forms the interdigitated electrodes.

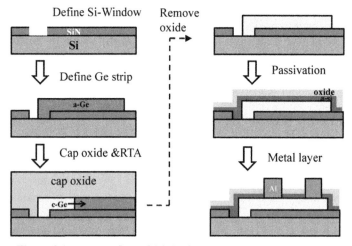

(a) Process flow of RMG-Ge method

(b) Process flow of Al/Ge/Al device

Figure 8.1   Process flow of fabricating Ge MSM photodetectors.

In Figure 8.2, the SAD pattern and HRTEM image of RMG Ge are displayed. It is clear that the SAD pattern in Figure 8.3a exhibits the monocrystalline phase of Ge. Another evidence of Ge crystallization can be found in Figure 8.2c, where micro-Raman spectroscopy was performed on samples with and without RTA. Without RTA, the Raman shift position and the FWHM are 270 and 42 ± 0.2 cm^{-1}, respectively. On the other hand, after RTA, those values shift to 299 and 4 ± 0.2 cm^{-1}, which are very close to the reported

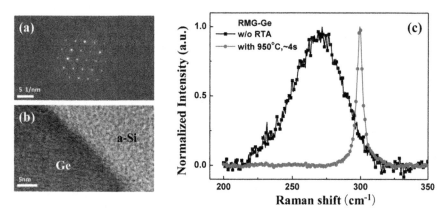

Figure 8.2   (a) SAD pattern, (b) HRTEM image, and (c) Raman shift of RMG-Ge.

data for Ge substrate ($\sim$298 and $\sim$3.2 cm^{-1}). In Figure 8.2b, the HRTEM image shows that RMG Ge is well passivated by a-Si, indicating that a high-quality interface between RMG Ge and a-Si is indeed achievable.

To further confirm that the RMG Ge is a single crystal, two test structures were examined; one is a Ge strip annealed with a seed window, while the other is not. Both samples have two Al contacts for measuring the IV characteristics. Figure 8.3 shows the I–V curves of the RMG-Ge MSM structures with or without the Si seed window, in the absence of a-Si passivation. The inset shows schematic of the device structure. In the case of devices with the seed window, the recrystallization of the melted Ge should replicate the crystalline phase of Si and proliferate to the strip end. On the other hand, without the Si-window, the annealed Ge inherently is poly-crystalline. Although the Schottky barrier exists at both the contacts of Al/monocrystalline Ge and Al/poly-crystalline Ge, the latter case introduces more interfacial states inside the Ge bandgap resulting from the grain boundary. Therefore, the carriers can tunnel through the Schottky barrier via the interfacial states. Such an effect can be observed from the measured I–V curves; for example, at a voltage of 0.5 V, the bias current from a Ge strip without the seed window is one order of magnitude larger than the one with the seed window.

Next, we investigate the effect of a-Si surface passivation. We prepared two devices with and without passivation, and their DC and PC were measured at the wavelength of 1.31 μm with an illumination power of 1.42 mW. The experimental results are summarized in Figure 8.4. Without passivation, the

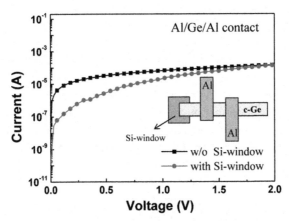

Figure 8.3 I–V characteristics of the Al/Ge/Al contact structure where the Ge strip is annealed with or without a Si seed window. No surface passivation is applied.

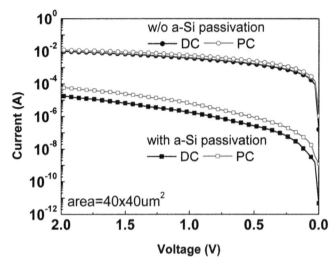

Figure 8.4   Measured DC and PC of the Al/Ge/Al photodetectors with and w/o a-Si passivation.

measured DC and PC are nearly the same ∼3.35 mA at the bias of 1 V, because the photocurrent is overwhelmed by the excessive surface leakages. On the other hand, with a-Si passivation, the measured DC and PC are 1.76 and 5.47 µA at 1 V, respectively. The DC is significantly reduced by three orders of magnitude compared to the non-passivated device, which demonstrates an effective surface passivation of Ge by a-Si. Note that the 30-nm thick a-Si layer deposited by PECVD covers only the top surface and the sidewalls of Ge strips. If a surrounded passivation technique is applied [15], the leakage current may further decrease.

As for the optical responsivity, we first analyze the quantum efficiency of light absorption for our MSM photodetector by FullWave (Rsoft Inc.) simulation software. Estimation of only 6.7% light absorption in the thin and arrayed Ge strips at 1.31 µm wavelength is obtained by excluding the reflected and transmitted optical powers. The low-photo-responsivity mainly results from metal reflection and small absorption of thin Ge strips. The optical responsivity, measured by taking the slope of PC versus absorbed optical power, is plotted as a function of bias voltage. The result is shown in Figure 8.5. Very low responsivity is observed for bias <1 V and starts to increase noticeably due to the photoconduction effect since the electric field is high enough to lower the Schottky barrier and drift the photocarriers swiftly from the absorption layer. At 2.0 V, optical responsivity around 0.4 A/W was measured.

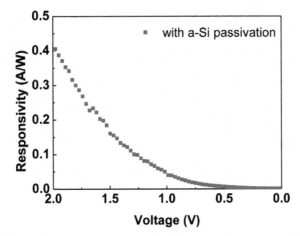

Figure 8.5 Optical responsivity of the Al/Ge/Al photodetectors with a-Si passivation.

8.3 Self-Assembled Microbonded Ge/Si Heterogeneous Structure

Germanium is one of the most essential semiconductors for state-of-the-art Si-based electronic and optoelectronic devices. High Ge mobility transistors [18] are fabricated on Si substrate, and high performance photodetectors [19] are implemented by Ge/Si heterojunctions for standard receiver modules. Because of lattice constant mismatch (4.2%) between Ge and Si, direct epitaxy Ge on Si usually requires critical processes and special tools [20, 21] to control the strain. On the other hand, growth temperature of epitaxy is often as high as 600–700° lasting for a long time (from 10 min to a few hours for a single run). However, higher thermal budget degrades the front-end process and also complicates the integration with other electronic devices. Another approach is wafer bonding, using a pre-epitaxial Ge wafer bonded to the Si wafer [22] and exfoliating the substrate. However, surface topography of the bonded surface usually affects the bonding yield. As demonstrated in the previous section, RMG [23, 24] is an effective route for integrating Ge on Si substrate. Nevertheless, it can't be employed for making a Ge/Si heterojunction.

In this section, a new idea of making a Ge/Si p–i–n heterojunction by exploiting surface tension (capillary force) to self-assemble RMG Ge on Si through a wet releasing process is presented. A self-assembled Ge/Si heterojunction waveguide photodiode is made with superior performance in terms of DC, on/off current ratio, operation bandwidth, and photoresponse.

8.3.1 Self-Assembled Microbonded Ge on Si by Surface Tension

Although RMG is an effective method for producing single-crystal Ge beams, direct contact between Ge and Si forming a heterojunction is prohibited because the Ge structure should be isolated from Si by a dielectric layer (except for the seed window) during the recrystallization process. Here, we propose that a Ge/Si heterojunction can be created by selectively removing the dielectric layer in a wet etching solution after RMG and using surface tension to bring Ge and Si contact together. Figure 8.6 schematically illustrates the idea. First, a RMG Ge beam enclosed by a silicon dioxide layer is immersed in a hydrofluoric acid solution for etching the oxide layer away. The Ge beam is thereby like a cantilever beam released in solution with one end anchored

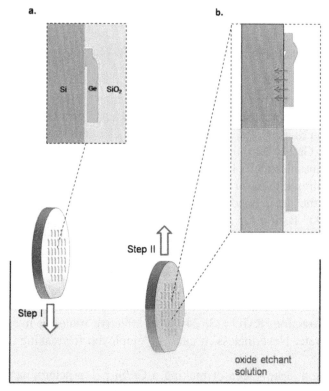

Figure 8.6   Self-assembly of Ge beams on Si surface. (a) Step I: the wafer is immersed in a solution-based BOE. The Ge beams are released by etching away the enclosed cap oxide. (b) Step II: the wafer is taken out from the solution. The released Ge beams adhere to the Si surface by surface tension [50].

on the substrate via the seed window. Next, once the wafer is leaving out from the etchant, because the wafer surface is hydrophobic, the trapped liquid underneath the Ge cantilever beam pulls the Ge beam adhering to the Si substrate by the surface tension. It is as if the Ge beam is bonded to the Si substrate. This phenomenon is commonly observed in the MEMS releasing process. Since the Ge beam is anchored on the substrate, the position is not altered after adhesion. In addition, the hydrophobic surface of Si and Ge also has the advantage of less water molecules trapped at the Ge/Si contact interface. Although the Ge beam is bonded on the Si surface, the Ge/Si contact majorly relies on the van der Waals' force. To increase the bonding strength, postannealing at a temperature of 400°C for 1 h in vacuum ambient is applied.

The occurrence of adhesion, as the Ge cantilever beam leaves from the solution, is dependent on the dimensions of the beam structure, Young's modulus of Ge, gap height, and surface energy of liquid trapped in the gap. A criterion to determine a critical cantilever beam length for adhesion is given by [25]

$$L > \sqrt[4]{\frac{3Et^3h^2}{8\gamma_s}},\tag{8.1}$$

where E is Young's modulus of Ge, t is the beam thickness, L is the beam length, h is the gap height between the beam and substrate, and γ_s is the surface energy of liquid. All these parameters are schematically described in Figure 8.7. For a Ge cantilever beam with a thickness of 0.3 μm and separated from the Si substrate by 50 nm, the beam length should at least be longer than 2 μm. Otherwise, the mechanical restoring force of the Ge cantilever beam can overcome the surface tension, leaving the beam standing above the Si surface without adhesion.

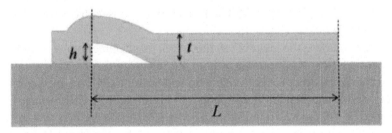

Figure 8.7 Schematic of a Ge beam (up layer) adhering to a Si substrate (bottom layer).

8.3.2 Device Schematics and Fabrication

The process flow of fabricating SAMB Ge pin waveguide photodetectors is described below. First, a 6-inch silicon-on-insulator (SOI) substrate with 500-nm device Si and 3-μm buried oxide is cleaned and n-type doped via ion implantation of phosphor with a concentration of 3×10^{18} cm^{-3}. The substrate is then deposited with a SiO$_2$ layer of 100 nm by LPCVD and leaves several seed windows contacting the bottom Si for the epitaxial growth. Then, e-beam evaporator is employed for depositing a-Si of 3 nm and a-Ge of 700 nm in thickness. Next, various Ge strips covering the seed windows are defined by ICP etching with Cl$_2$ and HBr as process gases. A thick cap oxide enclosed with Ge strips and the whole wafer was rapidly annealed under a temperature up to 950°C for 4 s. When cooling down, Ge was recrystallized from the seed window and proliferated to the end. All the dislocation defects resulting from Ge-Si lattice mismatch are trapped at the seed window step. Next, the wafer was immersed in BOE solution to release the Ge strips. These Ge strips are anchored at the seed window without flowing away and all oxide was etched by BOE etchant. When the wafer was taken out from the BOE solution, the Ge strips were self-aligned and bonded on the Si surface. This bonding comes from capillary of the liquid droplets leaving from the wafer surface. Once the surfaces come into contact, the bonding interface relies on van der Waal interaction to hold them together. This bonding phenomenon is often observed in MEMS devices [26]. In order to increase the bonding strength, an annealing process step is adopted. In this device, it was annealed at the temperature of 400°C for 1 h. Then, a back-end process continued to make a rib waveguide structure, which was patterned on the SOI layer underneath the Ge strip. The Si rib waveguide height and width are 100 nm and 3 μm, respectively. To decrease the surface leakage current from Ge, a thickness of 30 nm amorphous Si is deposited to passivate the Ge surface by PECVD. The anchor part of Ge strips was removed by dry etching to prevent the leakage current through the Si seed window. Finally, ILD coating, via etching, p-type Ge doping, nickel silicide, and metallization (TiN/AlSiCu/TiN/Ti) were consecutively executed to complete the devices.

Figure 8.8a shows the picture of the Ge/Si heterojunction waveguide photodetector. Optical wave propagating in the Si waveguide is evanescently coupled to the Ge absorber for converting optical power to PC collected by the designed p-Ge/i-Ge/n-Si photodiode shown in Figure 8.8b. A SEM image of the fabricated device is shown in Figure 8.8c, where the Ge beam is precisely aligned at the center of the Si waveguide. In Figure 8.8d, the HRTEM image displays the crystal orientations of Si and Ge atoms. There is a thin interfacial

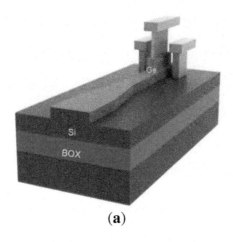

(a)

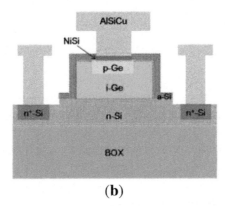

(b)

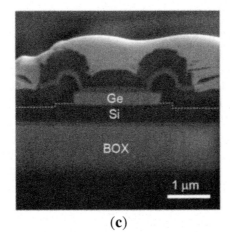

(c)

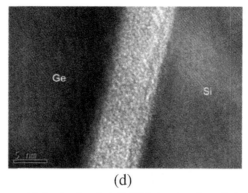

(d)

Figure 8.8   A self-assembled microbonded Ge/Si heterojunction photodiode. (a) Schematic of the device structure. The Ge optical absorber is bonded on the Si waveguide to form a vertical Ge/Si p–i–n photodiode. (b) Schematic of the device cross section. The Ge is p-type doped and the Si is n-type doped. (c) SEM of the fabricated device. The bonded 0.3 μm thick Ge beam is precisely aligned to the center of the Si waveguide. (d) HRTEM near the bonding interface. Crystal phase contrasts of Ge and Si are clearly observed and show the crystal orientations are well matched. A 7-nm interfacial layer exhibits between the Ge and Si [50].

layer between the Ge beam and Si waveguide, which is formed during the microbonding and postannealing process. This layer is measured to be 7 nm in thickness and is composed of Si, Ge, and O atoms, where oxygen accounts for 30% and the others are Si and Ge by investigating the material content via *in situ* energy EDS tool.

8.3.3 Results and Discussion

The I–V characteristics of this self-assembled Ge/Si heterojunction were investigated and the result is shown in Figure 8.9. A very low DC smaller than 7 nA was measured at a reversed bias voltage of –2 V. On the other hand, the forward bias current is about 5 mA at 2.5 V, corresponding to a six orders of magnitude of the on-off current ratio. These results are comparable to those of reported Ge/Si heterojunction waveguide photodetectors fabricated by heteroepitaxial growth. Photogenerated current tends to be saturated as the reverse bias voltage beyond –2 V, in case the intrinsic Ge layer is completely depleted.

In order to understand the electrical property of the interfacial layer, a temperature-dependent DC measurement was performed from 260 to 300 K, under different bias voltages. The DC can be generally characterized according to a band-band tunneling model [26].

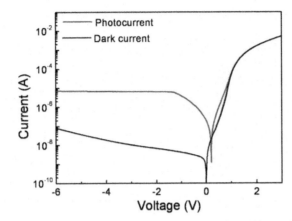

Figure 8.9 IV characteristics of a self-assembled microbonded Ge/Si heterojunction photodiode. The *red* and *black curves* are the PC and DC, respectively. The bonded area and thickness of the Ge beam are $15 \times 2 \, \mu m^2$ and $0.7 \, \mu m$. The DC is 7 nA at -2 V, corresponding to a current density of 23 mA/cm^2. The laser wavelength is 1310 nm [50].

$$I_{\text{Dark}} = BT^{3/2}e^{-E_a/kT}(e^{eV_a/2kT} - 1), \qquad (8.2)$$

where B is a proportionality constant, T is the temperature, E_a is the activation energy, and V_a is the bias voltage. By fitting the logarithm of measured DCs $\ln(I_{\text{Dark}}/T^{3/2})$ as a linear function of $1/kT$, the activation energy can be extracted with respect to different bias electrical fields shown in Figure 8.10. The activation energy decreases from 0.37 eV at 30 kV/cm to 0.19 eV at 92.3 kV/cm. Note that the activation energy 0.37 eV at a low bias condition is nearly half of the Ge bandgap, indicating that the Schockey-Read-Hall (SRH) recombination/generation via deep levels of trap states is responsible for the DC. With a stronger electric field, the activation energy reduces, and dark current increases. This could be explained by the effective Ge bandgap narrowing effect and larger band bending, which facilitates electrons and holes tunneling through the trap states. We emphasize that from the measured I–V curves, the DC is rather small in all biasing conditions, implying a low trap state density even in the presence of the interfacial layer.

The absorption length of a evanescently coupled a Ge/Si waveguide photodiode is estimated to be less than 5 μm. By using a 2D FDTD simulation carried out at 1310 nm wavelength. The Ge beam length of this device is designed to be 15 μm for complete absorption. The waveguide insertion loss, including fiber coupling loss, taper loss, and waveguide propagation loss, is calibrated to be 19 dB $\pm$ 0.5 dB. The large waveguide propagation loss is

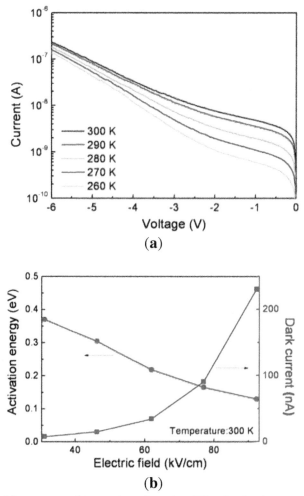

Figure 8.10 Measurement of temperature-dependent DC and activation energy. (a) Semilog plots of the DCs with respect to various temperatures ranging from 260 to 300 K. (b) Estimated activation energies versus electric fields and the corresponding DCs at room temperature [50].

mainly due to the rough waveguide sidewall of Si rib waveguides created by dry etching. By excluding the optical insertion loss, the device responsivity is estimated to be 0.66 and 0.70 (A/W) at the reverse bias of –4 and –6 V, respectively. The detailed characterization of photo-responsivity is displayed in Figure 8.11. Metal-induced absorption and scattering losses are mainly responsible for the lightly reduced responsivity. The red curve in Figure 8.11

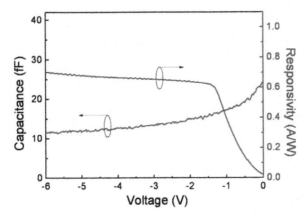

Figure 8.11 Characterization of photo-responsivity and capacitance. The blue curve is the measured photo-responsivity as a function of bias voltage by excluding the optical insertion loss, including fiber coupling, taper and waveguide propagation losses. The red curve is the measured total capacitance as a function of bias voltage, which is slightly larger than the calculated junction capacitance of 7 fF. The input optical power is 800 μW at 1310 nm wavelength [50].

represents the measured device capacitance with respect to the bias voltage. The capacitance is generally less than 15 fF and decreases slightly with a larger reverse bias, indicating that the Ge layer is completely depleted with a small junction capacitance.

High-frequency optical signal from an external modulator is coupled to the Si rib waveguide for measuring the device operation bandwidth. Figure 8.12 displays the measured 3 dB bandwidth ranging from 12 to 17 GHz, slightly increasing with the bias voltage from –4 to –6 V. Assuming the hole saturation velocity of Ge is 6×10^6 cm/s, the carrier transit time is thereby estimated to be 11.6 ps. On the other hand, the device RC delay time based on the measured junction capacitance and series resistance is calculated to be 7.7 ps. By considering both time constants, the analytically calculated device operation bandwidth is 17 GHz, agreeing well with the measured 3 dB bandwidth.

8.4 Self-Aligned Butt-Coupling of Ge and Si Waveguide

Integrated GeSi waveguide photodetectors had been investigated by many groups [27–32]. However, most devices are based on the evanescent coupling configuration; that is, the optical wave is coupled from the Si waveguide

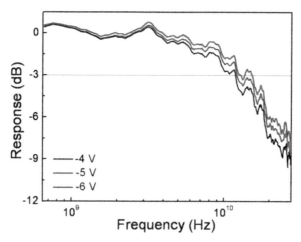

Figure 8.12   High-frequency measurement on photodiode bandwidth at 1310 nm wavelength. A Network Analyzer (E8361C, Agilent Technologies Inc.) is used to investigate the frequency response [50].

to a Ge detector on a different layer. This coupling approach is only efficient for sub-wavelength Si waveguides but not for large-core Si rib waveguides. Large-core Si rib waveguides are widely applied and are superior in low polarization-dependent loss and easy for optical fiber coupling. To have better optical coupling for those waveguides, the Ge detector should be butt-coupled to the Si waveguide core to have an efficient light absorption. Selective epitaxial growth in a trench could be a solution [33–36], but the process is critical and complex. However, no sample approach is presented to epitaxially grow single crystal Ge integrated with Si devices on the same planar structure.

In the previous section, we designed and fabricated a Ge/Si heterojunction pin waveguide photodetector through RMG in combination with a technique SAMB [39]. RMG has been demonstrated with capability of being implemented through standard CMOS fabrication technology [38]. The optical wave is evanescently coupled from a Si waveguide to a Ge absorber directly contacted on the waveguide. In this section, we further modify the SAMB process to deploy the Ge absorber butt-coupled to the Si waveguide for realizing a MSM PD.

Ge MSM PDs show great advantage in simple fabrication and high-speed detection [29]. However, Fermi level pinning is a serious issue that could lower SBH of the VB [40], resulting in a large leakage current. This issue becomes even severe when the PD is operated at a large bias voltage or under light illumination [41–43]. The Ge interfacial states may capture photogenerated carriers, leading to the carriers accumulated at the edges of metal contact where a strong image force is generated to further lower the SBH. This could cause device turn-on early during operation and could hinder the speed. Large bandgap materials, such as amorphous-Ge(a-Ge) [44], a-Si [45, 46], and silicon-carbon (Si:C) [47], inserted between the Ge and metal contact, can be utilized to suppress SBH lowering. Here, we use a-Si to modulate the SBH and passivate Ge surface to assure that the device can operate for high speed with a low leakage current.

8.4.1 Device Schematics and Fabrication

Figure 8.13 shows the device schematic of the butt-coupled Ge MSM PD. This long-strip single-crystalline Ge absorber is made by RMG [37] and subsequently self-assembled on a predefined silicon waveguide recess at the end. During the assembly process, it is inevitable to create a tiny gap between the Si and Ge parts. We then discuss the influence of the tiny gap on the optical coupling efficiency between the Si waveguide and Ge absorber in the following section. Two electrodes contact the Ge covered with an a-Si layer, and they are spaced by a distance of 900 nm to form a MSM junction to extract the PC.

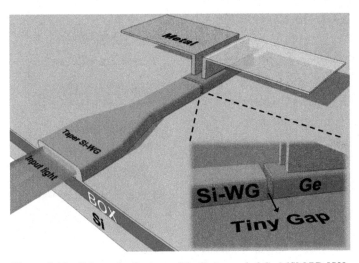

Figure 8.13 Schematic diagram of the butt-coupled Ge MSM PD [59].

The process flow of making the devices is illustrated in Figure 8.14. The device is fabricated on a 6-inch SOI wafer with 3 μm BOX and 340 nm Si layer. First, a 300-nm TEOS oxide was deposited at 600°C. This oxide layer was patterned as a hard mask to define a rib SOI waveguide, where the rib width is 2 μm and the rib height is 240 nm. The Si recess was simultaneously created with an etching depth of 240 nm during this process step. A 50-nm TEOS oxide was coated on the wafer except for several open windows exposed to the bottom Si as a seed. Subsequently, a 240 nm a-Ge film was deposited and then patterned with various long strips which are 2 μm in width and have a length varying from 5 to 10 μm. Each Ge strip covers the seed window as well as the recess step, leaving an overlapped Ge on the top of a Si waveguide by 500 nm. The top and bottom Ge sections are disconnected because of the large step height (350 nm TEOS oxide plus 240 nm Si recess depth). After deposition of 1 μm thick cap oxide, RTA at 950°C for 4 s was carried out to turn the a-Ge melting and recrystallizing from the seed window toward the end of the strips. Then, the device was immersed in BOE solution to release the Ge strips, where these strips were self-assembled onto the Si waveguide

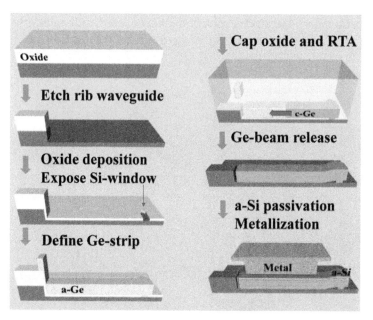

Figure 8.14   Process flow of making the butt-coupled Ge MSM PD (side view).

recess [39]. Meanwhile, the overlapped Ge originally covering the step was flushed away during the release process. Then, a 15-nm a-Si was deposited to passivate Ge surface and to modulate the SBH [29] of metal contact. Finally, ILD was deposited and a metallization process was applied to make the metal contacts and electrodes, which are stacked by AlSiCu/TiN/Ti multiple PVD system. Note that this process can be applied to a large core waveguide if it is fabricated on a thick SOI substrate.

Figure 8.15a shows a SEM image of a RMG Ge beam precisely assembled at the recess of a Si rib waveguide, where label 'C' marks the position of the seed window and label 'D' indicates an isolation trench near the Ge PD. A cross-section view of the Ge beam butt-coupled to the Si waveguide (from 'A' to 'B') is presented in Figure 8.15b. Prior Ge deposition, a thin TEOS oxide (50 nm) covers the whole wafer except the seed windows. Therefore, after RMG and Ge beam release, a gap emerges at the Ge and Si joint interface. This tiny gap was measured to be about 50 nm, corresponding with the thickness

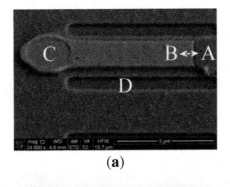

(a)

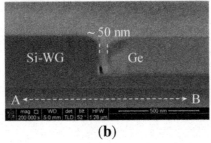

(b)

Figure 8.15 Fabricated and assembled Ge beam (5 μm in length) butt-coupled to a Si waveguide: (a) SEM image of the device (top view), (b) SEM image of the Ge beam at the recess step of the Si waveguide (cross-section view from 'A' to 'B'; [59]).

of the 50 nm TEOS oxide layer. This gap spacing can be precisely controlled by tuning the primed oxide layer thickness.

8.4.2 Optical Simulation on Coupling Efficiency

Figure 8.16a illustrates an optical wave propagating from a Si rib waveguide to a butt-coupled Ge absorber separated by a gap spacing d. Under the Ge absorber, there is a silicon slab layer (about 90 nm) remaining to form a composite Ge/Si waveguide structure. Figures 8.16b, c show the 2D optical mode profiles in the Si waveguide and Ge/Si absorber, respectively. The electric fields of these two modes match very well, indicating that the coupling efficiency should be high if the gap spacing can be reduced without introducing excess scattering loss. A 2D FDTD simulator (FullWAVE, Rsoft) is utilized to simulate the coupling efficiency varied by the gap spacing, and

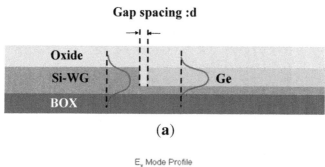

(a)

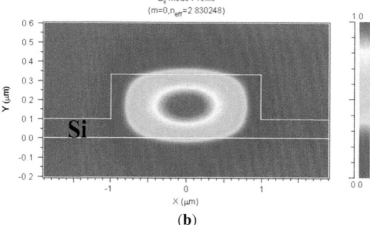

(b)

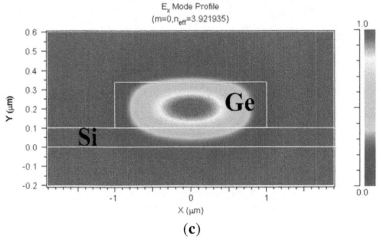

(c)

Figure 8.16 (a) Schematic illustration of optical wave propagating from a Si rib waveguide to a butt-coupled Ge/Si composite waveguide separated by a gap spacing d; simulated optical mode profiles in (b) Si rib waveguide and (c) Ge/Si waveguide [59].

the result is shown in Figure 8.17a. The dimensions of the Si waveguide, Ge/Si waveguide, and the etched recess in this simulation are based on the actual size of fabricated devices. The gap is filled with SiO_2 when the actual device is covered by ILD layer. Over 89% of the optical power coupled to the Ge/Si waveguide is achieved as the gap spacing is less than 50 nm. The 11% optical loss is contributed to light back reflection and scattering. If Ge absorption coefficient is considered, the simulated attenuation length is less than 2 μm.

In advance, the Ge absorber at the butt-coupling interface actually features a curved profile, which results from Ge melting during the RMG process. The surface tension force of Ge in liquid phase tends to modify the sharp corner to be rounded. In order to investigate whether the curved profile would impact the coupling efficiency, a series of simulations (2D-FDTD) is performed to examine the coupling efficiency versus the radius of curvature of the Ge interface. The result is shown in Figure 8.17b. Here, we suppose the gap spacing is 50 nm. The curved profile only introduces a small amount of scattering loss. For the real device, the radius of curvature is measured to be nearly 180 nm, by observing the SEM image in Figure 8.3b. Therefore, the coupling efficiency of the curved Ge interface is only 2% lower than the value with respect to a flat Ge-coupling interface.

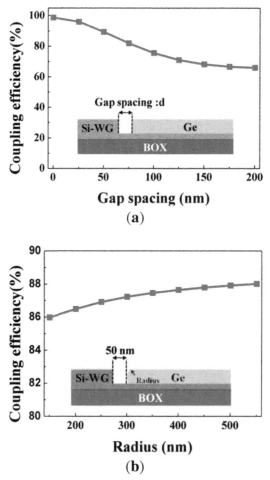

Figure 8.17   Simulated coupling efficiencies with respect to (a) gap spacing of Si-WG/Ge, and (b) radius of curvature of the Ge coupling interface [59].

8.4.3 Measurement Result and Discussion

Figure 8.18a displays the measured DCand PC at the wavelength of 1310 nm. To have a small coupling loss, a lensed fiber was used to couple light into the Si rib waveguide. The butt-coupled Ge MSM PD ($2 \times 10 \times 0.24 \ \mu m^3$) is characterized by a low DC before the device is turned on. Because a thin a-Si (15 nm) is inserted between Ti (metal) and Ge (semiconductor) to passivate the interfacial states and modulate the Schottky barrier height (SBH), the DC is significantly suppressed. The device structure and band

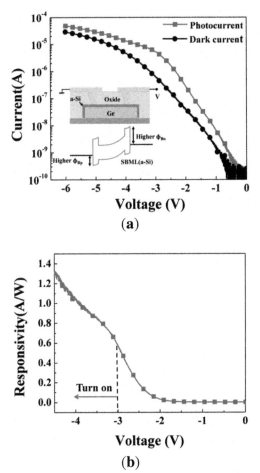

Figure 8.18 (a) DC and PC versus the bias voltage from 0 to –6 V and (b) responsivity versus the bias voltage from 0 to –4.5 V [59].

diagram of metal contact is illustrated in Figure 8.18a. To figure out the intrinsic photoresponsivity, the waveguide insertion loss, including both fiber coupling loss and waveguide propagation loss, is calibrated by measuring other straight rib waveguides with the same dimension but without the Ge PD. The waveguide power prior the Ge PD is estimated to be 8 μW. In Figure 8.18b, the photoresponsivity is estimated from 0.25 A/W at –2.6 V to 0.56 A/W at –3 V. In fact, the photoresponsivity increases linearly as the bias voltage exceeds –3 V, showing that the device is turned on and operated in photoconduction mode.

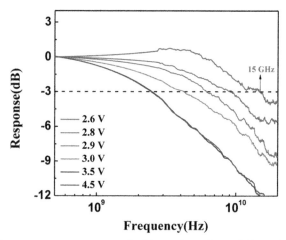

Figure 8.19   Frequency response versus the bias voltage from –2.6 to –4.5 V. The optical wavelength under test is 1310 nm [59].

The operation bandwidth of the butt-coupled Ge PD is investigated under different bias voltages. The result is shown in Figure 8.19. In as much as the Schottky barrier modulation layer, the MSM PD can be operated in the photodiode mode before the device is turned on. The operation bandwidth is mainly determined by the photocarrier transit time and the RC time constant. The measured 3 dB cut-off frequency is 15 GHz at –2.6 V. As the bias voltage is beyond 3 V, the device is turned on and operated in the photoconduction mode. Despite a large photoresponsivity, the 3 dB cut-off frequency is reduced to be 2 GHz, caused by a long photocarrier lifetime.

8.5 Germanium-Tin (GeSn) Photodetectors

Germanium-based optoelectronic devices integrated on silicon have been considered as a promising solution for silicon photonics to realize on-chip optical interconnects [48–50]. Ge is a quasi-direct bandgap semiconductor where the Γ-valley is close to the CB edge. The Ge active components such as light emitters and lasers have been demonstrated by several groups [51–53], via taking advantage of strained Ge or highly doped n-type Ge technologies. However, strained Ge exhibits a longer emission wavelength, resulting from the bandgap narrowing due to split of the heavy and light hole band at Γ point. This highlights the necessity of pursuing long wavelength detective material. One ideal candidate is GeSn alloy photodetectors [54, 55] since

GeSn is also a Group IV semiconductor that the process is more compatible with CMOS technology. However, higher concentration of Sn in GeSn alloy has been proven to be very challenging because of the lower solid solubility (1%) of Sn in Ge. One common approach to achieve a high concentration of Sn in Ge is through a low-temperature MBE system. However, the process temperature is too low (<400°C) to obtain a high-quality GeSn alloy. In this paper, a GeSn MSM photodetector is fabricated by the RMG method. This method has been demonstrated with ability to make a high-quality single crystal Ge device [56]. Furthermore, the fabrication process and cost are simpler and cheaper than conventional epitaxy, and more compatible with CMOS technology [57]. Based on this narrower bandgap material, photo-responsivity of GeSn is improved, compared to that of pure Ge material at longer wavelengths.

8.5.1 Device Schematics and Fabrication

Figure 8.20 shows the device structure of the GeSn photodetector on a SOI waveguide. The Sn elements concentrate at the end of the $Ge_{1-x}Sn_x$ strip during the RMG process, where the Sn distribution profile can be formulated according to the Scheil's equation [58]. A high concentration of Sn above the solid solubility is achievable near the end of the strip. A GeSn MSM contact configuration is employed with an amorphous silicon passivation layer, which suppresses the DC and modulates the SBH [59]. A laterally graded GeSn was

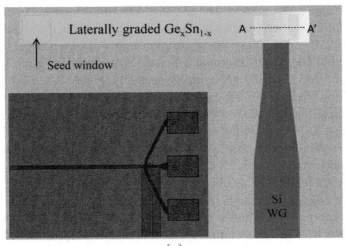

(a)

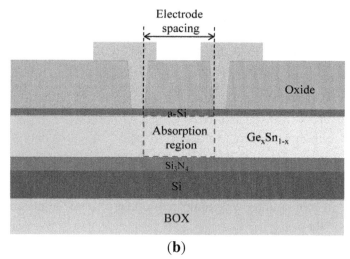

(b)

Figure 8.20   Schematic illustration on the GeSn MSM photodetector: (a) top view and (b) cross-sectional view (AA'). The inset shows a microscopy image of the fabricated device.

characterized by the EDS tool and showed a 2% Sn in the GeSn photodetector in this work. The GeSn strip dimension is 300 μm in length and 3 μm in width and intercepts a 3-μm wide Si waveguide.

An optical simulation (FullWAVE, Rsoft Ltd.) was carried out to analyze the absorption efficiency of the Si waveguide butt-coupled to the GeSn strip. The gap spacing was controlled by the thickness of a Si_3N_4 layer, and the optical wavelength was supposed to be 1310 and 1550 nm. Figure 8.21a shows the butt-coupling structure from a silicon waveguide of 260 nm thickness to a GeSn layer of 250 nm thickness. The absorption length (3 μm) is equal to the GeSn strip width. The power absorption efficiencies displayed in Figure 8.21b are 70 and 50% for 1310 and 1550 nm, respectively. Figure 8.22 illustrates the process flow of the GeSn waveguide MSM photodetector. Devices were fabricated on a SOI substrate with the Si layer of 260 nm and the BOX layer of 2 μm. First, a 300 nm TEOS oxide was deposited on top of the SOI, following by waveguide lithography and dry etching process. The slab height of the waveguide is 100 nm. A 50-nm Si_3N_4 layer was then deposited as an isolation layer for RMG, where only a few seed windows were left opened for directly contacting GeSn on the Si in the next process step. Second, a stacked structure of amorphous Ge/Sn/Ge (120/8/120 nm) was deposited layer by layer via an electron-gun evaporation system. The GeSn strips were defined by dry etching to cover the seed window. After the GeSn etching, a cap oxide

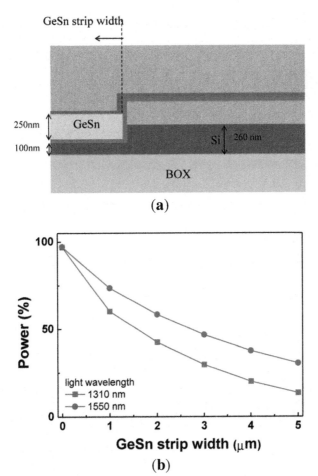

Figure 8.21 (a) The silicon waveguide butt-coupled to the GeSn strip; (b) absorption efficiency versus GeSn strip width.

layer with a thickness of 1 μm covered the whole structure and the wafer was annealed at 960°C for 4 s. During this step, the GeSn alloy melted and began to recrystallize from the seed window. Due to the low solubility of Sn in Ge, the Sn elements only distribute near the end of the strip. After the cap oxide was removed in a BOE solution, the GeSn strip was passivated with an amorphous silicon layer of 15 nm. Finally, the device was deposited with ILD layer and the metal lines (TiN/AlSiCu/TiN/Ti) were patterned and contacted on the device through via holes.

Cross section

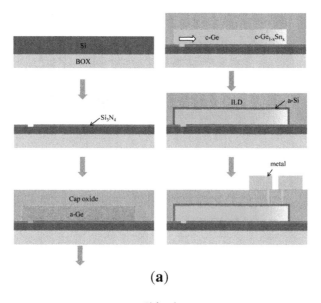

(**a**)

Side view

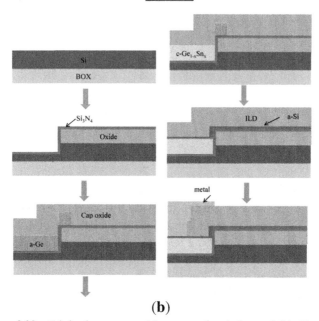

(**b**)

Figure 8.22 Fabrication process: (a) cross-sectional view and (b) side view.

8.5.2 Experimental Results

Through the RMG method, the GeSn strip re-crystallizes from the seed window, and all the threading dislocations are terminated at the corner of the step. Meanwhile, Sn concentration is redistributed along the strip during the crystallization process. In Figure 8.23(a, b), the Sn segregates at the end of the strip, where Sn precipitates and shows an interfacial plane. Near the region of Sn complete precipitation, the concentration of Sn atoms is actually beyond the solid solubility and dispersive within a range of a few microns. The photodetector is fabricated around 5 to 6 μm away from the strip end. The Sn concentration in this region is analyzed to be 2%, according to the EDX result. In addition, Figure 8.4c is a TEM image and x-ray SAD pattern of the graded $Ge_{1-x}Sn_x$ strip near the end, and it shows a high-quality GeSn crystallization.

Figure 8.24 shows the measured DC and PC of the GeSn MSM photo-detector, where the absorption region is $3 \times 3 \times 0.25$ μm^3. The optical wavelength is 1310 nm and the insertion loss is about 15 dB, including the fiber coupling loss as well as the waveguide loss. The metal line is made of

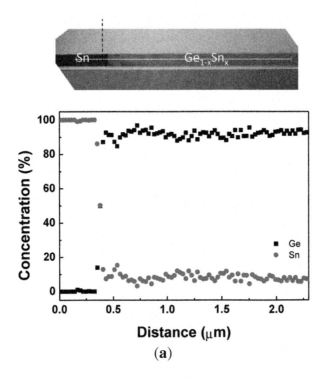

(a)

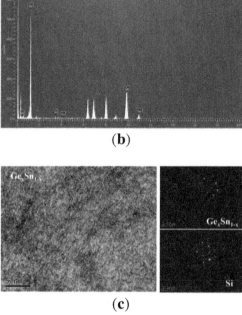

(b)

(c)

Figure 8.23   (a) The SEM image and EDX with line-scan mode at the end of the GeSn strip; (b) Energy Dispersive X-ray Analyzer (EDX) spectrum; and (c) HRTEM and diffraction pattern of $Ge_{1-x}Sn_x$ near the device region.

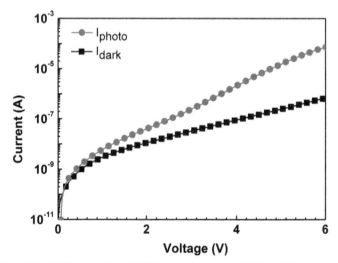

Figure 8.24   DC and PC of the GeSn MSM photodetector at 1310 nm. The electrode spacing is 3 μm.

TiN/AlSiCu/TiN/Ti, contacting the GeSn coated with an amorphous Si barrier layer of 15 nm. The DC is measured to be 3.7×10^{-7} A at a bias of 5.4 V. To verify a better absorption efficiency of GeSn at long wavelength, we compare the responsivity of Ge and GeSn photodetectors which have the same device structure. The result is shown in Figure 8.25. At 1310 nm, the responsivity of GeSn devices is not much different than that of Ge. However, at 1550 nm, the GeSn photodetectors exhibit a higher responsivity. It proves that GeSn has a larger absorption coefficient than Ge at long wavelength. The operation bandwidth of the butt-coupled GeSn waveguide photodetector is investigated. The bandwidth is shown in Figure 8.26. Because of the amorphous-Si barrier layer, device can be operated in the photodiode mode before the device is turned on. The bandwidth is determined by the transit time and the RC time constant. The measured 3 dB bandwidth is 6.2 GHz at 3.7 V for the device with electrode spacing of 2 μm. As the bias voltage is beyond 3.7 V, the bandwidth drops because the device is operated in the photoconductive mode. In summary, a GeSn MSM waveguide photodetector is implemented by the RMG method. The Sn concentration is 2%, which is beyond the solid solubility of Sn in Ge. The experimental results verify that the GeSn photodetector features better absorption efficiency at long wavelengths, compared with pure Ge devices.

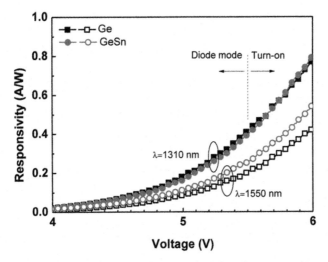

Figure 8.25 Responsivities of the Ge and GeSn MSM photodetectors at the wavelength of 1550 and 1310 nm. The bias voltage is 5.4 V. The electrode spacing is 3 μm.

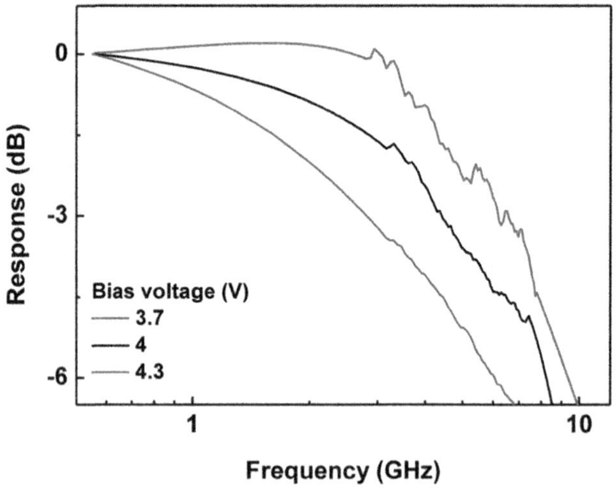

Figure 8.26   Bandwidth measurement of GeSn MSM photodetector. The electrode spacing is 2 μm.

8.6 Conclusion

The RMG method is demonstrated with ability to integrate heterogeneous semiconductors on Si substrate. We first show a single crystal, a-Si passivating germanium MSM photodetector made by this approach. Second, a novel SAMB technique is presented to form a Ge/Si heterojunction and applied for high-speed, high-performance Ge/Si waveguide photodiodes. This technique exploits capillary force to bond a Ge beam on a Si waveguide via a wet releasing process of etching SiO2, which is originally inserted between the Ge and Si devices for RMG. The device performance is excellent, which is comparable to most Ge/Si photodiodes. Then, a butt-coupled Ge MSM photodetectors is demonstrated by the SAMB technique to improve coupling efficiency between Si waveguide and Ge absorbers. Butt coupling is essential for thick Si waveguides coupled to Ge detectors but is difficult to be realized by traditional heterogeneous epitaxy. Finally, a graded GeSn alloy is accomplished by the RMG method to implement a GeSn MSM photodetector. The result shows better long-wavelength absorption efficiency than pure Ge devices. In summary, the RMG method is a wafer-level, cost-effective process applied for integrating a variety of semiconductors on Si substrate.

References

[1] H. C. Luan, et al., "High-quality Ge epilayers on Si with low threading-dislocation densities," Appl. Phys. Lett. **75** (19), 2909 (1999).

[2] Y. Liu, M. D. Deal, and J. D. Plummer, "High-quality single-crystal Ge on insulator by liquid-phase epitaxy on Si substrates," Appl. Phys. Lett. **87**, 14 (2009).

[3] S. Assefa, F. Xia, S. W. Bedell, Y. Zhang, T. Topuria, P. M. Rice, and Y. A. Vlasov, "CMOS-integrated high-speed MSM germanium waveguide photodetector," Opt. Express **18**, 5 (2010).

[4] S. Assefa, F. Xia, W. M. J. Green, C. L. Schow, A. V. Rylyakov and Y. A. Vlasov, "CMOS-Integrated Optical Receivers for On-Chip Interconnects," IEEE J. Sel. Top. Quantum Electron. **16** (5), 1376 (2010).

[5] S. Assefa, F. Xia and Y. A. Vlasov, "Reinventing germanium avalanche photodetector for nanophotonic on-chip optical interconnects," Nature. **464** (7285), 80 (2010).

[6] Y. M. Kang et al., "Monolithic germanium/silicon avalanche photodiodes with 340 GHz gain–bandwidth product," Nat. Photonics. **3** (1), 59 (2009).

[7] K. W. Ang, et al., "Low Thermal Budget Monolithic Integration of Evanescent-Coupled Ge-on-SOI Photodetector on Si CMOS Platform," IEEE J. Sel. Top. Quantum Electron. **16** (1), 106 (2010).

[8] J. Liu, J. Michel, W. Giziewicz, D. Pan, K. Wada, D. D. Cannon, S. Jongthammanurak, D. T. Danielson, and L. C. Kimerling, J. Chen, F. Ö. Ilday, F. X. Kärtner, J. Yasaitis, "High-performance, tensile-strained Ge p-i-n photodetectors on a Siplatform," Appl. Phys. Lett. **87**, 103501 (2005).

[9] J. Osmond, L. Vivien, J. M. Fédéli, Delphine M. M., P. Crozat, J. F. Damlencourt, Eric Cassan, Y. Lecunff, and S. Laval, "40 Gb/s surface-illuminated Ge-on-Si photodetectors," Appl. Phys. Lett. **95**, 151116 (2009).

[10] L. Chen and M. Lipson, "Ultra-low capacitance and high speed germanium photodetectors on silicon," Opt. Express **17**, 7901–7906 (2007).

[11] S. J. Koester, J. D. Schaub, G. Dehlinger, and J. O. Chu, "Germanium-on-SOI Infrared Detectors for Integrated Photonic Applications," IEEE J. Sel. Top. Quantum Electron. **12** (6), 1489 (2006).

[12] L. Chen, P. Dong, and M. Lipson, "High performance germanium photodetectors integrated on submicron silicon waveguides by low temperature wafer bonding," Opt. Express. **16** (15), 11513 (2008).

[13] K. W. Ang, M. B. Yu, S. Y. Zhu, K. T. Chua, G. Q. Lo and D. L. Kwong, "Novel NiGe MSM Photodetector Featuring Asymmetrical Schottky Barriers Using Sulfur Co-Implantation and Segregation," IEEE Trans. Electron Devices **45**, (9), 708–710 (1998).

[14] C. O. Chui, A. K. Okyay and K. C. Saraswat, "Effective Dark Current Suppression With Asymmetric MSM Photodetectors in Group IV Semiconductors," IEEE Photon. Technol. Lett. **15**, (11), (2003).

[15] M. Takenaka, K. Morii, M. Sugiyama, Y. Nakano and S. Takagi, "Dark current reduction of Ge photodetector by GeO_2 surface passivation and gas-phase doping," Opt. Express **20**, 8718–8725 (2012).

[16] M. Morse, O. Dosunmu, G. Sarid, and Y. Chetrit, "Performance of Ge-on-Si p-i-n Photodetectors for Standard Receiver Modules," IEEE Photon. Technol. Lett. **18**, 23 (2006).

[17] J. D. Hwang and E. H. Zhang, "Effect of a a-Si:H layer on reducing the dark current of 1310nm metal-germanium-metal photodetectors," Thin Solid films, **519**, 3819–3821(2011).

[18] C. C. Yeo, et al., "Electron Mobility Enhancement Using Ultrathin Pure Ge on Si Substrate," IEEE Electron Device Lett., 26 (2005).

[19] M. Morse, et al., "Performance of Ge-on-Si p-i-n Photodetectors for Standard Receiver Modules," IEEE Photon. Technol. Lett., 18 (2006).

[20] H. C. Luan, et al., "High quality Ge epilayers on Si with low threading-dislocation densities," App. Phys. Lett. 75, (1999).

[21] K. W. Ang, et al., "Low Thermal Budget Monolithic Integration of Evanescent-Coupled Ge-on-SOI Photodetector on Si CMOS Platform," IEEE J. Sel. Topic Quantum Electron. 16 (2010).

[22] A. J. Pitera, et al., "Coplanar Integrated of Lattice-Mismatched Semiconductors with Silicon by Wafer Bonding Ge/Si1-xGex/Si Virtual Substrate" J. Electrochem. Soc., 151, 7 (2004).

[23] Y. Liu, et al., "High-quality single-crystal Ge on insulator by liquid-phase epitaxy on Si substrates," App. Phys. Lett. 84, 2563–2565 (2004).

[24] J. Feng, et al., "Integration of Germanium-on-Insulator and Silicon MOSFETs on a Silicon Substrate," IEEE Electron Device Lett., 27 pp. 911–913 (2006).

[25] Mastrangelo, C. H. and Hsu, C. H., A simple experimental technique for the measurement of the work of adhesion of microstructures IEEE Solid-State Sensor and Actuator Workshop, 1992. 5th Technical Digest.

[26] Ang K. W., J. W. Ng, G. Q. Lo et al., Impact of field-enhanced band-traps-band tunneling on the dark current generation in germanium p-i-n photodetector, Appl. Phys. Lett. 94 (2009).

[27] M. Morse, O. Dosunmu, G. Sarid, and Y. Chetrit, "Performance of Ge-on-Si p-i-n photodetectors for standard receiver modules," *Ieee Photonics Technology Letters*, vol. 18, pp. 2442–2444, Nov–Dec 2006.

[28] Y. M. Kang, Y. Saado, M. Morse, M. J. Paniccia, J. C. Campbell, J. E. Bowers, and A. Pauchard, "Ge/Si Waveguide Avalanche Photodiodes on SOI Substrates for High Speed Communication," in *Sige, Ge, and Related Compounds 4: Materials, Processing, and Devices*. vol. 33, D. Harame, M. Ostling, G. Masini, T. Krishnamohan, S. Bedell, A. Reznicek, J. Boquet, Y. C. Yeo, M. Caymax, B. Tillack, S. Miyazaki, and S. Koester, Eds. Pennington: Electrochemical Soc Inc, 2010, pp. 757–764.

[29] S. Assefa, F. N. Xia, S. W. Bedell, Y. Zhang, T. Topuria, P. M. Rice, and Y. A. Vlasov, "CMOS-integrated high-speed MSM germanium waveguide photodetector," *Optics Express*, vol. 18, pp. 4986–4999, Mar 2010.

[30] C. T. DeRose, D. C. Trotter, W. A. Zortman, A. L. Starbuck, M. Fisher, M. R. Watts, and P. S. Davids, "Ultra compact 45 GHz CMOS compatible Germanium waveguide photodiode with low dark current," *Optics Express*, vol. 19, pp. 24897–24904, Dec 2011.

[31] S. R. Liao, N. N. Feng, D. Z. Feng, P. Dong, R. Shafiiha, C. C. Kung, H. Liang, W. Qian, Y. Liu, J. Fong, J. E. Cunningham, Y. Luo, and M. Asghari, "36 GHz submicron silicon waveguide germanium photodetector," *Optics Express*, vol. 19, pp. 10967–10972, May 2011.

[32] L. Chen, P. Dong, and M. Lipson, "High performance germanium photodetectors integrated on submicron silicon waveguides by low temperature wafer bonding," *Optics Express*, vol. 16, pp. 11513–11518, Jul 2008.

[33] D. Z. Feng, S. R. Liao, P. Dong, N. N. Feng, H. Liang, D. W. Zheng, C. C. Kung, J. Fong, R. Shafiiha, J. Cunningham, A. V. Krishnamoorthy, and M. Asghari, "High-speed Ge photodetector monolithically integrated with large cross-section silicon-on-insulator waveguide," *Applied Physics Letters*, vol. 95, p. 3, Dec 2009.

[34] L. Vivien, J. Osmond, J.-M. Fédéli, D. Marris-Morini, P. Crozat, J.-F. Damlencourt, E. Cassan, Y. Lecunff, S. Laval, and "42 GHz p.i.n Germanium photodetector integrated in a silicon-on-insulator waveguide," *Optics Eepress*, vol. 17, pp. 6252–6257, Apr 2009.

[35] S. Feng, Y. Geng, K. M. Lau, and a. A. W. Poon*, "Epitaxial III-V-on-silicon waveguide butt-coupled photodetectors," *Optics Letters*, vol. 37, pp. 4035–4037, Oct 2012.

[36] J. F. Liu, D. Ahn, C. Y. Hong, D. Pan, S. Jongthammanurak, M. Beals, L. C. Kimerling, J. Michel, A. T. Pomerene, C. Hill, M. Jaso, K. Y. Tu, Y. K. Chen, S. Patel, M. Rasras, A. White, D. M. Gill, and IEEE, *Waveguide integrated Ge p-i-n photodetectors on a silicon-on-insulator platform*. New York: IEEE, 2006.

[37] Y. C. Liu, M. D. Deal, and J. D. Plummer, "High-quality single-crystal Ge on insulator by liquid-phase epitaxy on Si substrates," *Applied Physics Letters*, vol. 84, pp. 2563–2565, Apr 2004.

[38] C.-K. Tseng, J.-D. Tian, W.-C. Hung, K.-N. Ku, C.-W. Tseng, Y.-S. Liu, N. Na, and M.-C. M. Lee, "Self-aligned microbonded Ge/Si PIN waveguide photodetectors," *Group IV Photonics (post-deadline session)*, Aug. 29–31 2012 Aug. 29–31.

[39] Solomon Assefa et al., "A 90 nm CMOS Integrated Nano-Photonics Technology for 25Gbps WDM Optical Communications Applications," *IEEE International Electron Devices Meeting (IEDM), (postdeadline session)*, Dec. 10–12 2012.

[40] T. Nishimura, K. Kita, and A. Toriumi, "Evidence for strong Fermi-level pinning due to metal-induced gap states at metal/germanium interface," *Applied Physics Letters*, vol. 91, Sep 17 2007.

[41] J. Burm and L. F. Eastman, "Low-frequency gain in MSM photodiodes due to charge accumulation and image force lowering," *Ieee Photonics Technology Letters*, vol. 8, pp. 113–115, Jan 1996.

[42] F. Xie, H. Lu, X. Q. Xiu, D. J. Chen, P. Han, R. Zhang, and Y. D. Zheng, "Low dark current and internal gain mechanism of GaN MSM photodetectors fabricated on bulk GaN substrate," *Solid-State Electronics*, vol. 57, pp. 39–42, Mar 2010.

[43] M. Klingenstein, J. Kuhl, J. Rosenzweig, C. Moglestue, A. Hulsmann, J. Schneider, and K. Kohler, "PHOTOCURRENT GAIN MECHANISMS IN METAL-SEMICONDUCTOR-METAL PHOTODETECTORS," *Solid-State Electronics*, vol. 37, pp. 333–340, Feb 1994.

[44] J. Oh, S. K. Banerjee, and J. C. Campbell, "Metal-germanium-metal photodetectors on heteroepitaxial Ge-On-Si with amorphous Ge Schottky barrier enhancement layers," *Ieee Photonics Technology Letters*, vol. 16, pp. 581–583, Feb 2004.

[45] B. Ciftcioglu, J. Zhang, R. Sobolewski, and H. Wu, "An 850-nm Normal-Incidence Germanium Metal-Semiconductor-Metal Photodetector With 13-GHz Bandwidth and 8-mu A Dark Current," *Ieee Photonics Technology Letters*, vol. 22, pp. 1850–1852, Dec 15 2010.

[46] J. D. Hwang and E. H. Zhang, "Effects of a a-Si:H layer on reducing the dark current of 1310 nm metal-germanium-metal photodetectors," *Thin Solid Films*, vol. 519, pp. 3819–3821, Mar 2011.

[47] A. Kah-Wee, L. Guo-Qiang, and K. Dim-Lee, "Germanium Photodetector Technologies for Optical Communication Applications," *Semiconductor Technologies, Jan Grym (Ed.), ISBN: 978-953-307-080-3, InTech, DOI: 10.5772/8572. Available from:* http://www.intechopen.com/books/semiconductor-technologies/germanium-photodetector-technologies-for-optical-communication-applications, 2010.

[48] S. Assefa, F. N. A. Xia, and Y. A. Vlasov, "Reinventing germanium avalanche photodetector for nanophotonic on-chip optical interconnects," *Nature*, vol. 464, pp. 80–U91, Mar 2010.

[49] T. Yin, R. Cohen, M. M. Morse, G. Sarid, Y. Chetrit, D. Rubin, and M. J. Paniccia, "31GHz Ge n-i-p waveguide photodetectors on Silicon-on-Insulator substrate," *Optics Express*, vol. 15, pp. 13965–13971, Oct 2007.

[50] C. K. Tseng, W. T. Chen, K. H. Chen, H. D. Liu, Y. M. Kang, N. Na, and M. C. M. Lee, "A self-assembled microbonded germanium/silicon heterojunction photodiode for 25 Gb/s high-speed optical interconnects," *Scientific Reports*, vol. 3, p. 6, Nov 2013.

[51] R. E. Camacho-Aguilera, Y. Cai, N. Patel, J. T. Bessette, M. Romagnoli, L. C. Kimerling, and J. Michel, "An electrically pumped germanium laser," *Optics Express*, vol. 20, pp. 11316–11320, May 2012.

[52] G. Shambat, S. L. Cheng, J. Lu, Y. Nishi, and J. Vuckovic, "Direct band Ge photoluminescence near 1.6 μm coupled to Ge-on-Si microdisk resonators," *Applied Physics Letters*, vol. 97, p. 3, Dec 2010.

[53] S. Saito, K. Oda, T. Takahama, K. Tani, and T. Mine, "Germanium fin light-emitting diode," *Applied Physics Letters*, vol. 99, p. 3, Dec 2011.

[54] D. L. Zhang, C. L. Xue, B. W. Cheng, S. J. Su, Z. Liu, X. Zhang, G. Z. Zhang, C. B. Li, and Q. M. Wang, "High-responsivity GeSn short-wave infrared p-i-n photodetectors," *Applied Physics Letters*, vol. 102, p. 4, Apr 2013.

[55] M. Oehme, M. Schmid, M. Kaschel, M. Gollhofer, D. Widmann, E. Kasper, and J. Schulze, "GeSn p-i-n detectors integrated on Si with up to 4% Sn," *Applied Physics Letters*, vol. 101, p. 4, Oct 2012.

[56] Y. C. Liu, M. D. Deal, and J. D. Plummer, "High-quality single-crystal Ge on insulator by liquid-phase epitaxy on Si substrates," *Applied Physics Letters*, vol. 84, pp. 2563–2565, Apr 2004.

[57] S. Assefa, S. Shank, W. Green, M. Khater, E. Kiewra, C. Reinholm, S. Kamlapurkar, A. Rylyakov, C. Schow, F. Horst, H. P. Pan, T. Topuria, P. Rice, D. M. Gill, J. Rosenberg, T. Barwicz, M. Yang, J. Proesel, J. Hofrichter, B. Offrein, X. X. Gu, W. Haensch, J. Ellis-Monaghan, Y. Vlasov, and Ieee, *A 90 nm CMOS Integrated Nano-Photonics Technology for 25 Gbps WDM Optical Communications Applications*. New York: Ieee, 2012.

[58] M. Kurosawa, Y. Tojo, R. Matsumura, T. Sadoh, and M. Miyao, "Single-crystalline laterally graded GeSn on insulator structures by segregation controlled rapid-melting growth," *Applied Physics Letters*, vol. 101, p. 4, Aug 2012.

[59] Wei-Ting Chen, C.-K. Tseng, Ku-Hung Chen, Han-Din Liu, Yimin Kang, Neil Na and Ming-Chang Lee, "self-Aligned Microbonded Germanium Metal-Semiconductor-Metal Photodetectors Butt-Coupled to Si Waveguides," *IEEE Journal of Selected Topics in Quantum Electronics*, 2013.

9

SiC Smart Photonic Waveguide Device for Data Processing

Chung-Lun Wu, Chih-Hsien Cheng, Yung-Hsiang Lin and Gong-Ru Lin

Graduate Institute of Photonics and Optoelectronics, and Department of Electrical Engineering, National Taiwan University, Taipei 10617, Taiwan, Republic of China

Abstract

The ultrafast non-linear optical Kerr switch with Si-QD doped in a-SiC (a-SiC:Si-QD) micro-ring resonator is demonstrated. The optical non-linearity of a-SiC can be significantly enhanced due to the enlarged oscillation strength of localized excitons in the Si-QD. The non-linear refractive index and TPA coefficient of a-SiC:Si-QD at 800 nm obtained from Z-scan measurements are 1.83×10^{-11} cm^2/W and 4.6×10^{-6} cm/W, respectively. Although the TPA effect is severed at 800 nm, it can be significantly suppressed by setting the operation wavelength at 1550 nm due to the small photon energy. Such a property is very important to analyze the non-linear Kerr switch at telecommunication wavelengths without interfering with the TPA and FCA. By injecting a pump-pulsed laser with peak power of 3 W into the a-SiC:Si-QD micro-ring resonator at resonance condition, the transmission spectrum is dynamically red-shifted to 0.07 nm due to the non-linear Kerr effect. By properly setting the probe wavelength at on-resonance and off-resonance of the a-SiC:Si-QD micro-ring resonator, the probe beam can be directly and inversely modulated by the injected pump source. Furthermore, the all-optical non-linear Kerr switch delivering NRZ-OOK data format with bit-rate of 12 Gbit/s has been successfully demonstrated by using the a-SiC:Si-QD micro-ring resonator.

Keywords: SiC, silicon QD, non-linear Kerr effect, all-optical modulation.

Green Photonics and Smart Photonics, 179–200.

9.1 Introduction

9.1.1 Historical Review of SiC-based Optoelectronic Devices

The non-stoichiometric SiC has been studied over decades because of its C/Si composition ratio detuned bandgap energy [1]. In particular, the SiC has been considered as the perfect matrix for high-power electronic devices due to its unique properties of high electron velocity [2] and large breakdown electric field [3]. When combining the features of controllable n- or p-type doping concentrations in SiC films [4], the SiC material has subsequently been considered as a potential candidate for optoelectronic devices. In the past decades, most researches related to the SiC-based optoelectronic devices have been focused on LEDs, SCs, and field-effect transistors. In view of previous works, the first report on the EL of p-i-n SiC LEDs varying from red to green color was observed by Kruangam et al. [5]. Later on, many studies on tuning the stoichiometry of SiC by changing its composition ratio were successively reported [6–8]. Tai et al. [9] have even utilized the Si-rich SiC films with buried silicon QDs and SiC-QDs as the active layers to improve the EQE of LEDs. The highest EQE of Si-rich-based LEDs could be obtained as $1.58 \times 10^{-1}\%$ [9]. Moreover, Gao et al. [10] also reported that the a-$Si_{1-x}C_x$:H n–i–p-based SC can be used as a semitransparent SC in an optical-transmittance modulator, but such a SiC SCS can only provide a conversion efficiency of less than 1%. Moreover, Cheng et al. [11] changed the thickness of i-SiC layer in the all Si-rich SiC-based SCs to promote the filling factor and conversion efficiency. The optimized thickness of i-SiC layer at 50 nm for all Si-rich SiC-based SCs enhanced its filling factor and conversion efficiency to 0.25 and 1.7%, respectively [11]. On the other hand, Estrada et al. [12] demonstrated that all the SiC-based thin-film transistors is improved its field effect mobility to 1.9×10^{-2} cm^2 V^{-1} s^{-1}. Nowadays, SiC materials are also utilized for fabricating high-temperature and high-power electronic devices to keep Si-based electronics sensitive under extreme environments by the superior thermal stability and chemical inertness of SiC [13, 14].

More recently, Si photonics have been developed for the application of optically interconnecting the electronic integrated chips because of bottle-necks under electrical transmission, which facilitates the development of hybrid photonic integrated chips with group IV semiconductor-based photonic and/or optoelectronic devices [15, 16]. More than that, the optoelectronic photonic devices based on other dielectric materials such as Si-rich SiO$_x$ and SiN$_x$ matrices have also emerged to meet some unique demands [17–19]. Accordingly, the unique non-linear optical properties of bulk SiC materials have been characterized for developing SiC-based waveguides and modulators [20, 21]. Neidermeier et al. [20] reported the experimental observation of second-order non-linear coefficients of SiC with different polytypes [20]. The second-order non-linear coefficient for 4H- and 6H-SiC materials are obtained as 18 and 24 pm/V, respectively [20]. Strait et al. [21] stated that the strong optical rectification effect of non-stoichiometric 6H–SiC could be effectively applied for the generation of terahertz radiation, and the ratio of $\chi_{zzz}^{(2)}/\chi_{zxx}^{(2)}$ for 6H–SiC was observed as -3 ± 2.6. Wu et al. [22] also calculated the second-order non-linear optical susceptibility of different SiC polytypes by employing the density functional theory. The $\chi_{xyz}^{(2)}$ for 3C–SiC and $\chi_{zzz}^{(2)}$ for 6H–SiC were simulated as 34.2 and 38.6 pm/V, respectively [22], except that there were few studies focused on the third-order non-linear optical properties of SiC materials [23, 24]. Only the analyses on damage threshold and optical non-linearity of bulk SiC measured by using the femtosecond laser were studied by DesAutels et al. [23]. Both the non-linear absorption coefficient and non-linear refractive index of the semi-insulating SiC have been determined as 6.4×10^{-2} cm/GW and 4.75×10^{-6} cm^2/GW, respectively. Ding's group also observed that the third-order non-linear optical susceptibility of the SiC material can be significantly modified with different nitrogen (N) doping concentrations [24]. By enlarging the N doping concentration to 2×10^{17} cm^{-1}, the real part of the third-order non-linear susceptibility can be improved to 6.16×10^{-13} esu [24]. Technically, there are many methods to measure the non-linear optical properties of materials, such as self-phase modulation [25], degenerate four-wave-mixing [26], non-linear interferometry [27], nearly degenerate three-wave mixing [28], ellipse rotation [29], beam distortion [30], and Z-scan measurements [31]. Among these aforementioned techniques, the Z-scan technology is commonly used due to its high sensitivity and simple equipment. Although the bulk SiC materials with their non-linear optical properties could be observed in previous reports, there are few works on studying the non-linear optical properties of nanoscaled SiC film and Si-QD doped in the a-SiC matrix.

9.1.2 Historical Review of Si-based All-Optical Switching with the Advantages of SiC-based Non-Linear Waveguide Applications

To achieve an ultra-high-speed communication system, the Si-based all-optical switching devices have been widely developed. Generally, the all-optical switching devices demonstrated by Si nanowires and Si-QD-based waveguides are based on the FCD, FCA and non-linear Kerr effect [32–39, 43]. Although the FCA cross-section in Si-QD is one-order of magnitude larger than the bulk Si [35], the free-carrier lifetime ($\sim$10 μs) is relatively long than bulk Si (1 ns). The free-carrier lifetime of the Si-QD doped in SiO_x matrix is dominated by the e–h pair recombination process but not influenced by the diffusion process (as in Si nanowire) [36]. By decreasing the Si-QD size, the carrier lifetime can be significantly enhanced up to $\sim$100 ns due to the quantum confinement effect [37]. However, the modulation bandwidth is still lower than the typical Si nanowire. Without the isolated electrical property of SiO_x host matrix, the free-carrier in the Si nanowire can be relaxed by diffusion processes [40, 41]. The modulation bandwidth of all-optical Si-based FCA modulator can be easily achieved $\sim$1 GHz [42]. Table 9.1 summarizes the performance specifications of Si-based modulators in recent years. Nevertheless, the FCA effect hardly meets the demand to obtain the ultrafast all-optical modulation speed. In this case, to achieve high-speed data transmission in the near future, the all-optical switching approach based on the non-linear optical Kerr effect must be considered for bit-rate improvement. The response time of the non-linear Kerr effect is around subpicosecond, which is much faster than the free-carrier lifetime in Si and Si-QD.

Recently, the all-optical Kerr switching in the Si-QD doped in SiO_x-based slot waveguide was realized [43]. By introducing the Si nanostructure into the SiO_2 host matrix, the optical non-linear property is much larger than that in stoichiometric SiO_2 [44–46]. As predicted by the quantum confinement effect, the enhanced dipole polarization contributed by the localized e–h pair in the Si-QD can significantly improve the three-order susceptibility [47]. However, the low refractive index of Si-QD doped in SiO_x (typical value of $\sim$1.8) leads to poor optical confinement for the channel waveguide geometry. The peak intensity is difficult to achieve when using the channel waveguide structure. Therefore, the slot waveguide structure was utilized to enhance the optical confinement for the SiO_x:Si-QD. In this case, the huge coupling loss between the slot waveguide and lensed fiber cannot be avoided due to the ultrasmall core size of the slot waveguide structure (typical core size of 50–100 nm). Concerning the Si nano-waveguide, the relatively low-bandgap energy of Si would result in a huge TPA by injecting the high-power pulse source at 1550 nm [48]. The TPA effect would decrease the optical power inside the waveguide, and the induced free-carrier can also degrade the non-linear Kerr effect. From this point of view, the Si-rich SiC is considered as a potential candidate. Because of the wide bandgap about 3.05 eV

Table 9.1 Historical review of Si-based modulators from recent years

Materials	Structure Type	Modulation Type/Operation Theory	Performance Specifications	Ref.
Si-QD doped in SiO$_x$ matrix	Rib waveguide	All-optical/FCA effect	6 dB/cm FCA loss at pump photon flux of $3 \times 10^{20}/cm^2 - s$	Navarro-Urrios et al. [34]
Si-QD doped in SiO$_x$ matrix	Rib waveguide	All-optical/FCA effect	FCA cross-section of 3.1×10^{-17} cm^2 at 1550 nm is one order of magnitude larger than bulk Si Modulation bandwidth increased to ~2 Mbit/s by shrinking Si-QD size to 1.7 nm	Wu et al. [36, 37]
Si	Channel waveguide with p-i-n diode	Electro-optical/FCA effect	Reduction of the free-carrier lifetime in Si nano-waveguide from 3 ns to 12.2 ps by applying a reverse bias across an integrated p-i-n diode.	Turner-Foster et al. [38]
Si	Micro-ring waveguide	All-optical/FCD effect	The optical transmission of the structure is modulated by more than 97% by use of control light pulse with energy as low as 40 pJ Response time of 450 ps	Almeida et al. [33]
Si$_3$N$_4$	Micro-ring waveguide	All-optical/Kerr effect	Non-linear refractive estimated of 2.4×10^{-15} cm^2/W by using non-linear Kerr switch Modulation bandwidth of 1 GHz is achieved	Ikeda et al. [39]
Si-QD doped in SiO$_x$ matrix	Micro-ring and slot waveguide	All-optical/Kerr effect	A modulation depth over 50% has been achieved for on-chip optical powers of the order of 100 mW Ultrafast modulation speed of 40 Gbit/s	Claps et al. [41]

and high-thermal stability [49], the SiC has the properties of low-absorption coefficient at 1550 nm and the material strength at high-power operation. Furthermore, the SiC is predicted to own the high-non-linear refractive index at telecommunication wavelengths. Based on these advantages, the optical non-linearity of SiC is expected to be improved by introducing the small Si-QD into the a-SiC host matrix.

9.1.3 Motivation and Chapter Content

In the first part of this chapter, the fabrication of Si-QD doped in a-SiC host matrix by using the hydrogen-free PECVD system is demonstrated. The atomic composition in the Si-QD doped in a-SiC is discussed. Furthermore, the non-linear optical property of the Si-QD doped in a-SiC film is analyzed by using the femtosecond Ti:Sapphire laser-based Z-scan measurement. In the second part, the ultrafast non-linear Kerr effect in the Si-QD doped a-SiC micro-ring resonator is investigated. The Si-QD doped a-SiC matrix all-optical switch is preliminarily demonstrated. The Si-QD doped a-SiC by low-temperature PECVD deposition process is a low-cost fabrication compared to the crystalline Si nano-waveguide fabricated by conventional CVD. Moreover, the non-linear optical properties of a-SiC can be tuned by adjusting the atomic composition or introducing the nanostructure into the host matrix. Such atomic composition variation in the SiC cannot be obtained in the single-crystallized SiC due to the fixed atomic composition and crystal structure. Figure. 9.1 illustrates the deposition of Si-QD doped in a-SiC matrix by using the low-temperature PECVD system. After the E-beam lithography and reactive ion etching process, the Si-QD doped in a-SiC micro-ring resonator can be demonstrated (refer to Figure 9.1). Based on the non-linear Kerr effect, the Si-QD doped in a-SiC micro-ring resonator is utilized to demonstrate the ultrafast all-optical switch. By using a pump-probe system, the continuous-wave (CW) probe signal can be directly and inversely modulated by the injected pump pulse. With the non-linear Kerr effect induced wavelength red-shift on the transfer function of the Si-QD doped in a-SiC micro-ring resonator, the non-linear refractive index at near-infrared wavelengths is also preliminarily determined. According to the ultrafast response of the non-linear Kerr effect, the refractive index inside the a-SiC ring cavity is dynamically modified by the input pump pulse. The transfer function of transmission properties then varies dynamically by the non-linear Kerr effect, thus providing a high-speed optical switch of up to 12 Gbit/s via the cross-wavelength amplitude modulation effect.

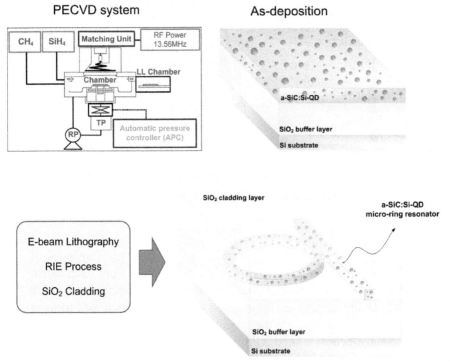

Figure 9.1 The fabrication process of a-SiC:Si-QD micro-ring resonator.

9.2 Structural Properties of Amorphous Si-Rich SiC

9.2.1 Composition of Amorphous Si-Rich SiC

The amorphous Si-rich SiC film was deposited on the Si wafer by using the PECVD with a mixed gaseous recipe of 10% diluted SiH_4 and pure CH_4. The fluence ratio (R_{SiC}) defined as the CH4 to total gas flow is set at 0.5 to deposit amorphous Si-rich SiC film. The C/Si composition ratios of these non-stoichiometric SiC films were determined by using XPS. By employing the XPS analysis, the atomics concentration is measured as 64.3% and 37.1% for Si and C elements, respectively. As compared by the stoichiometric SiC, the amorphous Si-rich SiC has the excessive concentration of 37.2% with RSiC of 50%. The C/Si composition ratio of the amorphous Si-rich SiC film grown with the R_{SiC} = 0.5 is 0.42. Under low temperature and weak RF plasma deposition, the SiH_4 molecules is more easily decomposed compared with the CH_4 molecules due to the lower dissociation energy of the SiH_4 molecules (75.6 kcal/mol) [50]. Under high molecule density, each reactant molecule

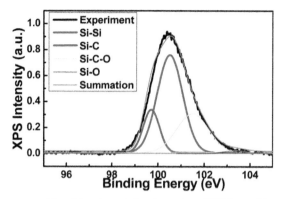

Figure 9.2 The compositional bonds and the dependent XPS intensities within the $Si_{(2p)}$ spectrum.

obtains insufficient energy from the plasma, and the decomposing rates of SiH_4 and CH_4 molecules cannot significantly distinguish from each other. Therefore, the Si-rich condition of SiC is easily obtained under the growth of $R_{SiC} = 0.5$. In the meantime, the oxygen content in Si-rich SiC films is also maintained as 8.6% to retain the quality of the PECVD grown SiC film. The $Si_{(2p)}$ orbital electron related XPS spectra of the Si-rich SiC film grown with $R_{SiC} = 0.5$ is shown in Figure 9.2. The detected XPS spectra indicates the significant phase change of SiC films by fitting the $Si_{(2p)}$ orbital electron-related XPS spectra with four separated Gaussian components. The binding energies of the decomposed peaks are 99.7, 100.5, 101.5 and 103.35 eV, which are attributed to the Si–Si bonds, Si–C bonds, C–Si–O bonds, and Si–O bonds, respectively. The presence of Si-Si bond in Si-rich SiC films indicates that the Si-QDs existed in the Si-rich SiC film.

9.2.2 Optical Non-Linearity of Amorphous Si-Rich SiC

The Z-scan technology has shown its potential for the analyses of non-linear refractive index and absorption coefficient due to its high-sensitivity and simplified architecture. By fitting the transmittance variation around the focal point of the Z-scan system, the optical non-linear properties of the a-SiC can be determined. Figure 9.3 shows the configuration of the single-beam Z-scan experiment including both open- and closed-aperture experiments. The pumping laser source is outputted from the femtosecond Ti:sapphire laser at wavelength of 800 nm. The pulsewidth and repetition rate are 80 fs and 80 MHz, respectively. The minimum beam diameter of the focused pump beam is ~20 μm, and the excitation intensity is varied by moving the sample

(a)

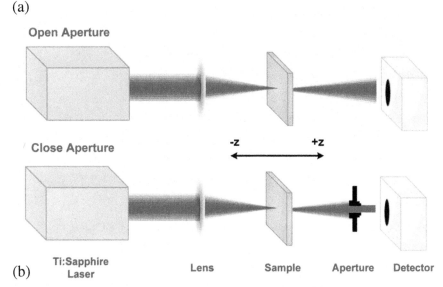

(b)

Figure 9.3 The scheme diagram of the open- (a) and closed-aperture (b) Z-scan measurement.

along the z-axis with a motorized translational stage. The far-field transmitted light passing through the aperture with its beam intensity is recorded by a balanced photodetector. For the open-aperture Z-scan analysis, the transmitted beam through the sample is not shut by the aperture. In that case, the non-linear absorption (TPA or saturated absorption) can be observed in the open-aperture Z-scan analysis. In contrast, a closed-aperture Z-scan only detects the on-axis part of the divergent and diffracted beam by placing an aperture on the z-axis.

The Z-scan traces for the a-SiC films with the R_{SiC} of 0.5 are shown in Figure 9.4. For the open-aperture Z-scan analysis, the transmittance near the focal point is decreased significantly. It implies that the two-photon absorption is observed in the Si-QD doped in a-SiC matrix at a wavelength of 800 nm. The characterization of the intensity-dependent absorption for a-SiC film can be performed [51], and the non-linear absorption coefficient of Si-rich SiC at 800 nm is $\sim 4.6 \times 10^{-6}$ cm/W determined by the open-aperture configuration. To further extract the non-linear refractive index of the Si-rich SiC, the close/open Z-scan trace shown in Figure 9.4 is obtained by dividing the closed-aperture Z-scan trace with the open-aperture Z-scan trace, and the non-linear refractive index can be fitted with the theoretical transmittance function [46, 52]:

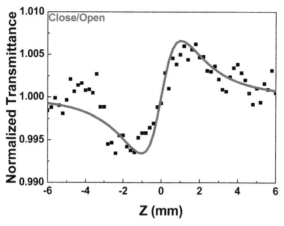

Figure 9.4   The close/open Z-scan transmittance ratio of the Si-rich SiC films.

$$T(z, \Delta\Phi) = 1 - \frac{4\Delta\Phi z/z_0}{(1 + (z/z_0)^2)(9 + (z/z_0)^2)}$$

$$= 1 - \frac{4n_2 k_0 I_{\text{peak}} L z/z_0}{(1 + (z/z_0)^2)(9 + (z/z_0)^2)}, \qquad (9.1)$$

where $\Delta\Phi$, n_2 and k_0 denote the phase shift, the non-linear refractive index of the a-SiC film, and the wave number, respectively. After numerical simulation, the non-linear refractive index of a-SiC film is obtained as 1.83×10^{-11} cm^2/W for a-SiC films grown with the R_{SiC} of 0.5. This result clearly elucidates that the excessive Si nanocluster could effectively provide large non-linear optical properties in the Si-rich SiC than the crystalline SiC.

9.3 All Optical Switching in Si-QD-Doped in a-SiC Micro-Ring Resonator

9.3.1 Fabrication of Si-QD-Doped in a-SiC Micro-Ring Resonator

Prior to determining the geometric structure of a-SiC-based ring resonator, the refractive index of Si-QD-doped Si-rich SiC film was obtained by fitting the reflection spectrum of a-SiC:Si-QD film deposited on Si substrate. The refractive index of a-SiC:Si-QD is calculated as ~2.63 at wavelength of 1.5 μm. For single-mode operation in a-SiC:Si-QD-based channel waveguide, the width and height of a-SiC:Si-QD core layer are set as 600 and 300 nm, respectively. For fabricating the a-SiC:Si-QD-based micro-ring channel waveguide, the a-SiC:Si-QD film is deposited on the Si substrate, which is covered with 3-μm-thick SiO_2 by thermal oxidation. Subsequently, the electron beam lithography is performed to define the ring and bus waveguide. The width of the waveguide is set as 600 nm, and the gap between the ring and bus waveguide is set as 300 nm. The diameter of the ring resonator is 300 μm. To enhance the coupling efficiency between the waveguide facet and lensed fiber, the inverse taper structure is introduced into the waveguide design [53]. The width of the inverse taper is varied from 200 to 600 nm within a length of 200 μm. After the E-beam lithography, the Cr layer with 80 nm is deposited on the a-SiC:Si-QD film by using E-gun evaporation. Afterward, the Cr hard mask is transferred on the a-SiC:Si-QD film with the lift-off process. Then, the reactive-ion-etching (RIE) process with a recipe of $CF_4 + O_2$ is used to remove the unpatterned a-SiC:Si-QD and form the a-SiC:Si-QD-based micro-ring resonator. After removing the Cr mask, a 2-μm-thick SiO_2 upper-cladding layer is deposited by using the PECVD. Finally, both end-facets are cleaved and polished to minimize its coupling loss of smaller than 3 dB/facet. The polished waveguide cross-section is shown in Figure 9.5a, which reveals that the waveguide facet is very smooth, and the interface between the waveguide core and cladding can be clearly observed. The top-view images of the inverse taper and micro-ring resonator are demonstrated in Figure 9.5b, c, respectively. The diameter of the ring resonator is set as 300 μm, which is larger than ever reported. This is because the bending loss contributed by the ring waveguide is expected to be minimized.

9.3.2 Operation of SiC Ring Resonator

With the presence of the micro-ring resonator, the output transmission power is modified with the dark comb-like throughput transfer function on the notched transmission spectrum, as shown in Figure 9.6. The transmission spectrum of a-SiC:Si-QD micro-ring resonator shows the duel-modes at long wavelengths. This originates from the TE_0 and TM_0 modes in the a-SiC:Si-QD micro-ring resonator. The extension of transmission dip of the TM_0 mode is lower than that of the TE_0 mode due to the lower power coupling between the ring and bus waveguides of the TM_0 mode. As evidence, by comparing the mode tails between TE_0 and TM_0 modes, the evanescent wave of the TE_0 mode spreads more significantly than the TM_0 mode and results in the high-extension transmission dip. In order to obtain the optical property of the a-SiC:Si-QD-based micro-ring resonator, the normalized transmission spectrum of a-SiC:Si-QD micro-ring resonator can be simulated by using the equation as shown in Rabus [54].

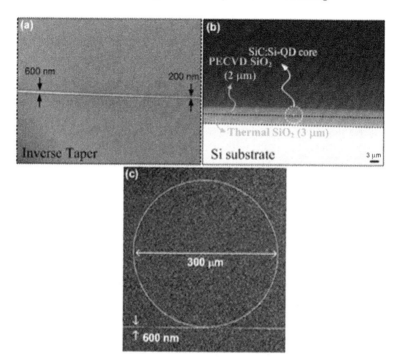

Figure 9.5 (a) The top view image of inverse taper structure of a-SiC:Si-QD-based waveguide. (b) The cross-sectional image of polished a-SiC:Si-QD-based waveguide, and (c) The SEM top view image of a-SiC:Si-QD-based micro-ring resonator.

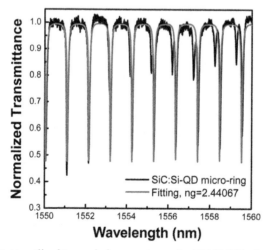

Figure 9.6 The normalized transmission spectrum of a-SiC:Si-QD micro-ring resonator.

The operation principle for demonstrating the non-linear Kerr switch is illustrated in Figure 9.7. A CW optical probe signal and a high-power optical pump data-stream at optical telecommunication wavelengths are concurrently coupled into the Si-rich SiC channel micro-ring waveguide resonator. By injecting the high-power optical pulsed data-stream at any resonance dip of the transmission spectrum of a-SiC:Si-QD micro-ring resonator, the transmission resonance dips can be dynamically red-shifted with the presence of an intensive pump due to the non-linear Kerr effect. Such non-linear Kerr effect causes an increased refractive index in the a-SiC:Si-QD micro-ring resonator and results in the red-shift of the notched resonant dip away from its original wavelength. As a result, the transmittance at probe wavelength is dynamically influenced by the non-linear Kerr effect. By properly selecting the probe wavelength around the resonance dip of the a-SiC:Si-QD micro-ring resonator, the probe beam can be directly or inversely modulated by the pump pulse due to the non-linear Kerr effect. In more detail, if the wavelength of the probe beam is set at the resonance dip of the micro-ring resonator without pump pulse injection, the transmittance of the probe beam reduces from its initial condition. When the pump pulse is injected into the ring resonator, the

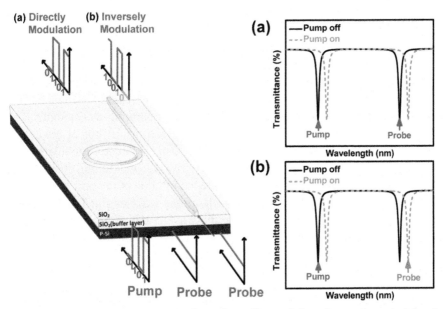

Figure 9.7 The schematic diagram of non-linear Kerr switch and operation principle of (a) direct modulation and (b) inverse modulation.

resonance dip is red-shifted due to the non-linear Kerr effect. In that case, the transmittance of the probe beam is increased accordingly. Then, the probe beam can be directly modulated by the pump pulse, as shown in Figure 9.7a. On the contrary, if the wavelength of the probe beam is slightly adjusted away from the resonance dip (longer than the resonant wavelength), the transmittance of the probe beam is increased without the pump pulse modulation. Once the pump pulse is injecting into the a-SiC:Si-QD micro-ring resonator, the resonance dip is dynamically red-shifted to the probe wavelength due to the non-linear Kerr effect. That is, the transmittance of the probe beam is instantly decreased when the pump pulse is introduced into the ring cavity, resulting in the inverse modulation of the probe beam (referred to in Figure 9.7b).

The non-linear Kerr switch is characterized by a pump-probe analysis. An external modulation method is used to generate a high-power optical pump pulse. An electrical pulse with a duration of 83 ps and a repetition rate of 12 MHz was employed to modulate a tunable laser through a Mach–Zehnder modulator. The pump pulse was amplified by using an EDFA to obtain a peak power of 3 W. To inject the pump/probe beam into the a-SiC:Si-QD micro-ring resonator, two beams are combined by using a 50/50 coupler injected into the waveguide via a lensed fiber. Moreover, the modulated probe beam and pump pulse are collected from another waveguide facet by using the lensed fiber. In order to analyze the modulated probe signal without the contribution of the pump pulse, the optical bandpass filter is utilized to eliminate the pump pulse. Subsequently, the modulated probe signal is detected by using a high-speed photodetector, and the modulated probe trace is displayed by using the digital sampling oscilloscope. Figure 9.8 shows the time-domain traces of a single bit shape for the modulated probe signals with different operating wavelengths. Firstly, when selecting the probe wavelength at the resonance dips of the a-SiC:Si-QD micro-ring resonator, the probe beam can be directly modulated by the original optical pump data-stream and with the maximum positive modulated amplitude. Moreover, the modulated amplitude of the probe beam is gradually decreased with the red shift of the probe wavelength. The probe beam becomes inversely modulated when the wavelength of probe beam is shifted. In comparison with the time-domain trace of the high-power pump pulse, the converted probe and the inverted probe perfectly match the bit shape to the optical pump pulse without distortion.

To realize the practical NRZ-OOK modulation by using the a-SiC:Si-QD micro-ring resonator, the optical pump source is encoded by an AWG with the NRZ-OOK data format. The bit-rate of the pump signal is 12 Gbit/s and the time-domain trace is shown in Figure 9.9a. Similar to the previous

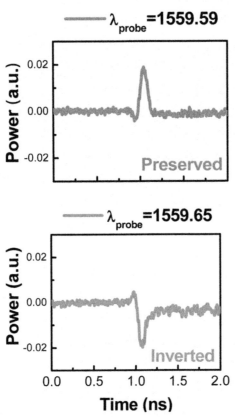

Figure 9.8 The preserved and inverted single bit shapes of the modulated probe signals at different operating wavelengths.

experimental condition, the wavelength of the pump signal is selected at resonance dip of the micro-ring resonator of 1551.08 nm to induce the non-linear Kerr effect. As expected, when setting the probe wavelength at on-resonance (λ_{probe} = 1559.59 nm) and off-resonance (λ_{probe} = 1559.65 nm) in the adjacent transmission dip of the micro-ring resonator, the probe beam can be directly and inversely modulated accordingly. The corresponding modulated probe signal traces with preserved and inverted NRZ-OOK data-stream are shown in Figures 9.9b, c, respectively. Moreover, there is no signal distortion by comparing the pump data-stream with the modulated probe signal traces, indicating that the 12 Gbit/s NRZ-OOK data-stream can be modulated by using the a-SiC:Si-QD micro-ring resonator. Furthermore, modulation on the probe signal is dominated by the red-shift on the resonance dip of the

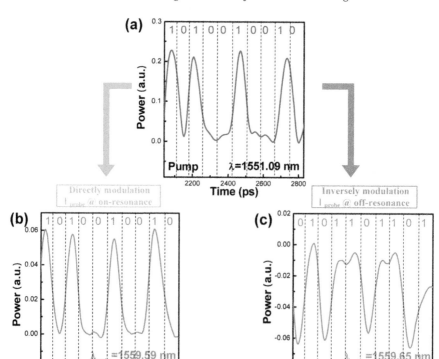

Figure 9.9   The time-domain traces of the 12 Gbit/s NRZ-OOK data-streams measured from the ports of (a) the optical pump input at λ = 1551.08 nm, (b) the probe output at on-resonant dip wavelength of 1559.59 nm, and (c) the probe output at off-resonant dip wavelength of 1559.65 nm.

a-SiC:Si-QD micro-ring resonator, which is solely contributed by the non-linear Kerr effect but not by the TPA or FCA effect.

9.4 Conclusion

The optical non-linearity of a-SiC is enhanced by doping the Si-QD into the host matrix by using the PECVD. Based on the Z-scan measurement, the non-linear refractive index and two-photon absorption coefficient at 800 nm of a-SiC:Si-QD are 1.83×10^{-11} cm^2/W and 4.6×10^{-6} cm/W, respectively. Furthermore, the a-SiC:Si-QD is utilized to fabricate the micro-ring resonator to demonstrate the optical non-linear Kerr switch at $\sim$1550 nm. The fabricated micro-ring waveguide resonator is obtained with Q = 22,117 and the transmittance drop at 1551.08 nm of 60%. By injecting the high-power pump pulse of 3 W at 1551.08 nm into the micro-ring resonator, the refractive index change induced by the non-linear Kerr effect is utilized to modify the transmittance of the micro-ring resonator. Therefore, the CW probe beam can be modulated by the pump pulse due to the non-linear Kerr effect. Finally, the all-optical modulation in the a-SiC:Si-QD micro-ring resonator with 12 Gbit/s NRZ-OOK data format has been successively realized in the present work. The modulated probe signal traces show no distortion in comparison with the pumping data-stream, indicating that the modulation process is dominated by the non-linear Kerr effect without the contribution of TPA and FCA effects. Such a-SiC:Si-QD-based ultrafast all-optical Kerr switch shows great potential for the non-linear optical waveguide-based data processing applications.

References

[1] Demichelisa F, Crovinia G, Pirria CF, Tressoa E, Gallonib R, Rizzolib R, Summonteb C, Zignanic F, Ravad P, Madane A. The influence of hydrogen dilution on the optoelectronic and structural properties of hydrogenated amorphous silicon carbide films. Philos. Mag. B. 1994; 69: 377–386. DOI: 10.1080/01418639408240116.

[2] Bhatnagar M, Baliga BJ. Comparison of 6H-SiC, 3C-SiC, and Si for power devices. IEEE Trans. Electron Devices. 1996; 40: 645–655. DOI: 10.1109/16.199372.

[3] Wan J, Capano MA, Melloch MR. Formation of low resistivity ohmic contacts to n-type 3C-SiC. Solid State Electron. 2002; 46: 1227–1230. DOI: 10.1016/S0038-1101(02)00013-8.

[4] Demichelis F, Pirri CG, Tresso E. Influence of doping on the structural and optoelectronic properties of amorphous and microcrystalline silicon carbide. J. Appl. Phys. 1992; 72: 1327–1333. DOI: 10.1063/1.351742.

[5] Kruangam D, Endo T, Wei PG, Nonomura S, Okamoto H, Hamakawa Y. A study of visible-light injection-electroluminescence in a-SiC/p-i-n diode. J. Non-Cryst. Solid. 1985; 77–78: 1429–1432. DOI: 10.1016/0022-3093(85)90924-X.

[6] Lee CT, Tsai LH, Lin YH, Lin GR. A chemical vapor deposited silicon rich silicon carbide p-n junction based thin-film photovoltaic solar cell. ECS J. Solid State Sci. Technol. 2012; 1: Q144–Q148. DOI: 10.1149/2.005301jss.

[7] Lin GR, Lo TC, Tsai LH, Pai YH, Cheng CH, Wu CI, Wang PS. Finite silicon atom diffusion induced size limitation on self-assembled silicon quantum dots in silicon-rich silicon carbide. J. Electrochem. Soc. 2011; 159: K35–K41. DOI: 10.1149/2.014202jes.

[8] Lo TC, Tsai LH, Cheng CH, Wang PS, Pai YH, Wu CI, Lin GR. Self-aggregated Si quantum dots in amorphous Si-rich SiC. J. Non-Cryst. Solids. 2012; 358: 2126–2129. DOI: 10.1016/j.jnoncrysol.2012.01.013.

[9] Tai HY, Cheng CH, Lin GR. Blue-green light emission from Si and SiC quantum dots co-doped Si-rich SiC p-i-n junction diode. IEEE J. Sel. Top. Quantum Electron. 2014; 20: 8200507. DOI: 10.1109/JSTQE.2013.2291701.

[10] Tawada Y, Kondo M, Okamoto H, Hamakawa Y. Hydrogenated amorphous silicon carbide as a window material for high efficiency a-Si solar cells. Sol. Energy Mater. 1982; 6, 299–315. DOI: 10.1016/0165-1633(82)90036-3.

[11] Cheng CH, Lin YH, Chang JH, Wu CI, Lin GR. Semi-transparent Si-rich SiC p-i-n photovoltaic solar cell grown by hydrogen-free PECVD. RSC Adv. 2014; 4: 18397–18405. DOI: 10.1039/C3RA41173G.

[12] Estrada M, Cerdeira A, Resendiz L, García R, Iñiguez B, Marsal LF, Pallares J. Amorphous silicon carbide TFTs. Solid State Electron. 2006; 50: 460–467. DOI: 10.1016/j.sse.2006.03.001.

[13] Huran J, Hrubcin L, Kobzev AP, Liday J. Properties of amorphous silicon carbide films prepared by PECVD. Vaccum. 1996; 47: 1223–1225. DOI: 10.1016/0042-207X(96)00128-5.

[14] Cooper JA, Melloch MR, Singh R, Agarwal A, Palmour JW. Status and prospects for SiC power MOSFETs. IEEE Trans. Electron Devices. 2002; 49: 658–664. DOI: 10.1109/16.992876.

[15] Corte F, Rao S, Coppola G, Summonte C. Electro-optical modulation at 1550 nm in an as-deposited hydrogenated amorphous silicon p-i-n waveguiding device. Opt. Express. 2011; 19: 9421–9451. DOI: 10.1364/OE.19.002941.

[16] Irrera F, Lemmi F, Palma F. Transient behavior of adjustable threshold a-Si:H/a-SiC:H three-color detector. IEEE Trans. Electron Devices. 1997; 44: 1410–1416. DOI: 10.1109/16.622595.

[17] Lin GR, Lin CJ, Yu KC. Time-resolved photoluminescence and capacitance-voltage analysis of the neutral vacancy defect in silicon implanted SiO2 on silicon substrate. J. Appl. Phys. 2004; 96: 3025–3027. DOI: 10.1063/1.1775041.

[18] Lin CD, Cheng CH, Lin YH, Wu CL, Pai YH, Lin GR. Comparing retention and recombination of electrically injected carriers in Si quantum dots embedded in Si-rich SiN_x films. Appl. Phys. Lett. 2011; 99: 243501. DOI:10.1063/1.3663530.

[19] Wu CL, Lin GR. Gain and emission cross section analysis of wavelength-tunable Si-nc incorporated Si-rich SiO_x waveguide amplifier. IEEE J. Quantum Electron. 2011; 47: 1230–1237. DOI: 10.1109/JQE.2011.2161459.

[20] Niedermeier S, Schillinger H, Sauerbrey R, Adolph B, Bechstedt F. Second-harmonic generation in silicon carbide polytypes. Appl. Phys. Lett. 1999; 75: 618–620. DOI: 10.1063/1.124459.

[21] Strait J H, George PA, Dawlaty J, Shivaraman S, Chandrashekhar M, Rana F, Spencer MG. Emission of terahertz radiation from SiC. Appl. Phys. Lett. 2009; 95: 051912. DOI: 10.1063/1.3194152.

[22] Wu IJ, Guo GY. Second-harmonic generation and linear electro-optical coefficient of SiC polysytpes and nanotubes. Phys. Rev. B. 2008; 78: 035447. DOI: 10.1103/PhysRevB.78.035447.

[23] DesAutels GL, Brewer C, Walker M, Juhl S, Finet M, Ristich S, Whitaker M, Powers P. Femtosecond laser damage threshold and nonlinear characterization in bulk transparent SiC materials. J. Opt. Soc. Am. B. 2008; 25: 60–66. DOI: 10.1364/JOSAB.25.000060.

[24] Ding JJ, Wang YC, Zhou H, Chen Q, Qian SX, Feng ZC, Lu WJ. Nonlinear optical properties and ultrafast dynamics of undoped and doped bulk SiC. Chin. Phys. Lett. 2010; 27: 124202. DOI: 10.1088/0256-307X/27/12/124202.

[25] Yoshio E, Teraoka M, Broaddus DH, Kita T, Tsukazaki A, Kawasaki M, Gaeta AL, Yamada H. Self-phase modulation at visible wavelengths in nonlinear ZnO channel waveguides. Appl. Phys. Lett. 2010; 97: 071105. DOI: 10.1063/1.3480422.

[26] Friberg SR, Smith PW. Nonlinear optical glasses for ultrafast optical switches. IEEE J. Quantum Electron. 1987; QE-23: 2089–2094. DOI: 10.1109/JQE.1987.1073278.

[27] Weber MJ, Milam D, Smith WL. Nonlinear refractive index of glasses and crystals. Opt. Eng. 1978; 17: 463–469. DOI: 10.1117/12.7972266.

[28] Adair R, Chase LL, Payne SA. Nonlinear refractive index measurement of glasses using three-wave frequency mixing. J. Opt. Soc. Am. B. 1987; 4: 875–881. DOI: 10.1364/JOSAB.4.000875.

[29] Owyoung A. Ellipse rotations studies in laser host materials. IEEE J. Quantum Elect. 1973; QE-9: 1064–1069. DOI: 10.1109/JQE.1973.1077417.

[30] Williams WE, Soileau MJ, Stryland E. Optical switching and n_2 measurements in CS_2. Opt. Commun. 1984; 50: 256–260. DOI: 10.1016/0030-4018(84)90328-6.

[31] Sheik-bahae M, Said AA, Stryland E. High-sensitivity, single-beam n_2 measurements. Opt. Lett. 1989; 14: 955–957. DOI: 10.1364/OL.14.000955.

[32] Soref RA, Bennett BR. Electrooptical effects in silicon. IEEE J. Quantum Electron. 1987; QE-23: 123–129. DOI: 10.1109/JQE.1987.1073206.

[33] Almeida VR, Barrios CA, Panepucci RR, Lipson M, Foster MA, Ouzounov DG, Gaeta AL. All-optical switching on a silicon chip. Opt. Lett. 2004; 29: 2867–2869. DOI: 10.1364/OL.29.002867.

[34] Navarro-Urrios D, Pitanti A, Daldosso N, Gourbilleau F, Rizk R, Pucker G, Pavesi L. Quantification of the carrier absorption losses in Si-nanocrystal rich rib waveguides at 1.54 μm. Appl. Phys. Lett. 2008; 92: 051101. DOI: 10.1063/1.2840181.

[35] Kekatpure R, Brongersma M. Quantification of free-carrier absorption in silicon nanocrystals with an optical microcavity. Nano Lett. 2008; 8: 3787–3793. DOI: 10.1021/nl8021016.

[36] Wu CL, Su SP, Lin GR. All-optical data inverter based on free-carrier absorption induced cross-gain modulation in Si quantum dot doped SiO_x waveguide. IEEE J. Sel. Topics Quantum Electron. 2014; 20: 8200909. DOI: 10.1109/JSTQE.2013.2292885.

[37] Wu CL, Su SP, Lin GR. All-optical modulation based on silicon quantum dot doped SiO_x:Si-QD waveguide. Laser Photonics Rev. 2014; 8: 766–776. DOI: 10.1002/lpor.201400024.

[38] Turner-Foster AC, Foster MA, Levy JS, Poitras CB, Salem R, Gaeta AL, Lipson M. Ultrashort free-carrier lifetime in low-loss silicon nanowaveguides. Opt. Express. 2010; 18: 3582–3591. DOI: 10.1364/OE.18.003582.

[39] Ikeda K, Saperstein RE, Alic N, Fainman Y. Thermal and Kerr nonlinear properties of plasma-deposited silicon nitride/silicon dioxide waveguides. Opt. Express. 2008; 16: 12987–12994. DOI: 10.1364/OE.16.012987.

[40] Linnros J. Carrier lifetime measurements using free carrier absorption transients. II. Lifetime mapping and effects of surface recombination. J. Appl. Phys. 1998; 84: 284–291. DOI: 10.1063/1.368025.

[41] Claps R, Raghunathan V, Dimitropoulos D, Jalali B. Influence of non-linear absorption on Raman amplification in silicon waveguides. Opt. Express. 2004; 12: 2774–2780. DOI: 10.1364/OPEX.12.002774.

[42] Reed GT, Mashanovich G, Gardes FY, Thomson DJ. Silicon optical modulators. Nature Photon. 2010; 4: 518–526. DOI: 10.1038/nphoton.2010.179.

[43] Martinez A, Blasco J, Sanchis P, Galan JV, Garcia-Ruperez J, Jordana E, Gautier P, Lebour Y, Hernandez S, Spano R, Guider R, Daldosso N, Garrido B, Fedeli JM, Pavesi L, Marti J. Ultrafast all-optical switching in a silicon-nanocrystal-based silicon slot waveguide at telecom wavelengths. Nano Lett. 2010; 34: 1506–1511. DOI: 10.1021/nl9041017.

[44] Prakash GV, Cazzanelli M, Gaburro Z, Pavesi L, Iacona F, Franzo G, Priolo F. Nonlinear optical properties of silicon nanocrystals grown by plasma-enhanced chemical vapor deposition. J. Appl. Phys. 2002; 91: 4607–4610. DOI: 10.1063/1.1456241.

[45] Hernández S, Pellegrino P, Martínez A, Lebour Y, Garrido B, Spano R, Cazzanelli M, Daldosso N, Pavesi L, Jordana E, Fedeli JM. Linear and nonlinear optical properties of Si nanocrystals in SiO_2 deposited by plasma-enhanced chemical-vapor deposition. J. Appl. Phys. 2008; 103: 064309. DOI: 10.1063/1.2896454.

[46] Spano R, Daldosso N, Cazzanelli M, Ferraioli L, Tartara L, Yu J, Degiorgio V, Jordana E, Fedeli JM, Pavesi L. Bound electronic and free carrier nonlinearities in silicon nanocrystals at 1550 nm. Opt. Express. 2009; 17: 3941–3950. DOI: 10.1364/OE.17.003941.

[47] Chen R, Lin DL, Mendoza B. Enhancement of the third-order nonlinear optical susceptibility in Si quantum wires. Phys. Rev. B. 1993; 48: 11879–11882. DOI: 10.1103/PhysRevB.48.11879.

[48] Manolatou C, Lipson M. All-optical silicon modulators based on carrier injection by two-photon absorption. J. Lightwave Technol. 2006; 24:1433–1439. DOI: 10.1109/JLT.2005.863326.

[49] Lefèvre J, Costantini JM, Esnouf S, Petite G. Thermal stability of irradiation-induced point defects in cubic silicon carbide. J. App. Phys. 2009; 106: 083509. DOI: 10.1063/1.3245397.

[50] Cheng CH, Wu CL, Chen CC, Tsai LH, Lin YH, Lin GR. Si-rich SiC light-emitting diodes with buried Si quantum dots. IEEE Photonics J. 2012; 4:1762–1775. DOI: 10.1109/JPHOT.2012.2215917.

[51] Kamaraju N, Kumar S, Kim YA, Hayashi T, Muramatsu H, Endo M, Sood AK. Double walled carbon nanotubes as ultrafast optical switches. Appl. Phys. Lett. 2009; 95: 081106. DOI: 10.1063/1.3213396.

[52] Sheik-Bahae M, Said AA, Wei TH, Hagan DJ, Stryland EWV. Sensitive measurement of optical nonlinearities using a single beam. IEEE J. Quantum Electron. 1990; 26: 760–769. DOI: 10.1109/3.53394.

[53] Almeida VR, Panepucci RR, Lipson M. Nanotaper for compact mode conversion. Opt. Lett. 2003; 28: 1302–1304. DOI: 10.1364/OL.28.001302.

[54] Rabus DG. Integrated Ring Resonators, Ring Resonators: Theory and Modeling. 2007. Springer, 3–40, Berlin Heidelberg. DOI: 10.1007/978-3-540-68788-7.

10

Application of Ion Beam Technology in the Synthesis of ZnO Nanostructures

Liang-Chiun Chao

Department of Electronic and Computer Engineering,
National Taiwan University of Science and Technology,
Taipei, Taiwan 106

Abstract

Ion beam technology is a fundamental tool in the semiconductor industry that it has found numerous vital applications. Broad beam ion sources are commonly used in surface cleaning, material deposition, and dopant implantation, while field emitter ion sources are used in nanofabrication. ZnO is a wide band gap semiconductor material. With a direct band gap of 3.34 eV and an exciton binding energy of 60 meV, ZnO is a plausible candidate for UV light emitting application. Besides, ZnO also exhibits a rich variety of nanostructures. In this chapter, the research work done by our group over the past few years is summarized, which includes the development of broad beam capillaritron ion sources, formation of metallic zinc nanocones, ZnO QDs on ion beam textured substrates, and ZnO nanowires by thermal oxidation. Special emphasis is given to the characterization of ZnO nanowires prepared by thermal oxidation. Our results show that by utilizing ion beam processing to minimize impurities in zinc, ZnO nanowires obtained by thermal oxidation show excellent PL property, fast photoresponse, and exceptional high photoconductive gain.

Keywords: Ion beam, ZnO nanowires, thermal oxidation.

Green Photonics and Smart Photonics, 201–218.

10.1 Introduction

Ion beam is a vital tool in semiconductor processing that it has been widely used in thin film preparation, dopant implantation, and surface cleaning. Ion sources are commonly divided into either broad beam or field emitter type ion sources. Among the broad beam ion sources, capillaritron [1] is an ion source with a simple architecture. First reported by Mahoney in 1980, a capillaritron is composed of a capillary anode and a grounded cathode positioned at close proximity to the anode. The anode is a quartz capillary tubing coated with a conductive material. As high voltage is applied to the capillary and working gas is fed through the capillary, the electric field between the anode and the cathode results in formation of plasma via electron impact ionization.

One of the most important field emitter type ion source is LMIS. Generated from the development of ion thrusters in the 1960s, LMIS has sparked a new era in ion beam applications [2], mainly due to its high brightness and small virtual source size. The brightness of LMIS is almost three orders of magnitude higher than that of dualplasmatron. Due to the fact that ions are much heavier than electrons, electrostatic lenses must be used in an ion optical column. For a DC voltage ion optical column with axially symmetric electrostatic lenses without space charge, nor gauze type lenses, the spherical aberration is always positive [3]. FIB work station with optimized ion optical column utilizing LMIS is capable to deliver a target current density of ~ 2 A/cm^2 [2]. This high-target current density enables a wide variety of applications, including nanofabrication and integrated circuit repair. Most of the metal elements can be converted into liquid metal or liquid alloy ion source form. However, because of corrosion and vapor pressure issues, only Ga, In, and Au-Si are commercially available [2, 4, 5]. The high brightness and the small virtual source size properties make LMIS the only choice for FIB applications. Up to now, broad beam plasma ion sources are not suitable for FIB work station due to their much larger virtual source sizes, lower brightness [6], and the always positive spherical aberration of the ion optical column [3].

Low-dimensional metal oxide semiconductor materials are of great value for next-generation optoelectronic devices, for which the surface effect may dominate their overall physical properties. Due to its high surface-to-volume ratio, low-dimensional semiconductor materials with exceptional high sensitivity have been demonstrated. ZnO is one of the materials that exhibit a rich variety of nanostructures, including QDs, nanodisk, nanotube, nanobelt, nanorods, and nanowires. In this chapter, the capillaritron ion source developed by our group is presented, which is then used for the preparation of ZnO nanostructures and the properties of ZnO nanowires are characterized using PL. PL which is a powerful characterization tool that the defect states of ZnO can be unambiguously identified. Our results show that by reducing the contaminants in zinc, ZnO nanowires obtained by ion beam processing and thermal oxidation exhibit excellent PL and photosensing properties.

10.2 Capillaritron Ion Source

Figure 10.1 shows a photo of a capillaritron ion source developed by our group. This ion source is designed following Mahoney's [1] approach that a metallic capillary acts as the anode and is positioned at close proximity to a cathode, which is ground to earth. As working gas is passed through this capillary, the electric field between the cathode and the anode capillary results in the formation of plasma that is maintained by electron bombardment. Ions are then accelerated through the opening on the cathode. This ion source is capable of operating with argon, oxygen and nitrogen. During operation, the vacuum chamber is backed-filled with the working gas up to several mTorr. The outer shell diameter of the ion source is 32 mm so that it easily fits through a CF2.75" flange. The capillary is composed of a nickel alloy tubing with an orifice diameter of 100 μm. This ion source is capable of operating with beam energy from 0.5 up to 15 keV.

Figure 10.2 shows target current density of the capillaritron ion source operated with argon at discharge currents of 70 and 350 μA. The target current was measured by a positioning Farady cup at 35 mm downstream of the ion source. From Figure 10.2, this ion source is capable of delivering a target

Figure 10.1 The capillaritron ion source developed at Taiwan Tech. This ion source is mounted on a CF2.75" flange with a SHV-20 feedthrough.

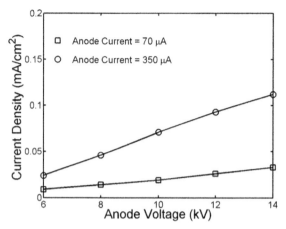

Figure 10.2 Target current density of a capillariton ion source operated with argon at discharge currents of 70 and 350 μA.

current density of 100 μA/cm². It is interesting to note that the target current density is roughly 10,000 times lower than that of LMIS, even though the total discharge current is 100 times higher. LMIS usually operates with a total current of 3–5 μA. This indicates that the virtual source size of capillaritron is much larger than that of LMIS.

10.3 ZnO Nanostructures

ZnO nanowire or nanorods have been prepared by various procedures, including chemical vapor deposition, thermal evaporation, and vapor phase transport. The growth mechanisms are either VLS mechanism or VS mechanism [7], with or without the use of catalysts. Vapor phase transport method utilizes mixed ZnO and graphite powder with a molar ratio close to 1:1. After heating the mixed powder at ∼1000°C, ZnO is decomposed by graphite that results in the formation of zinc and zinc suboxide [8]. Both zinc and zinc suboxide are carried by the carrier gas to a low temperature zone where ZnO nanostructures are deposited. A common procedure to prepare ZnO nanostrcuture based devices is to separate these ZnO nanostructures from the substrate by sonication. Separated ZnO nanostructures are then spread on substrates which have certain predefined patterns. To complete the device fabrication, e-beam lithography is utilized to make contacts from the predefined patterns to the nanostructures [7, 9].

Figure 10.3 shows some of the ZnO nanostructures that we have synthesized over the past few years. Figure 10.3a shows ZnO nanodonuts, which are obtained by heating mixed ZnO, graphite, and Er_2O_3 powder at 1000°C [10]. Depth-dependent ESCA analysis indicates that these nanodonuts are actually porous [11]. Figure 10.3b shows ZnO nanowires prepared by thermal oxidation of metallic zinc, which will be discussed in detail later. Figure 10.3c shows ZnO QDs deposited on ion beam textured substrates [12]. The ion beam textured substrates are important in controlling QD size and density distribution. Figure 10.3d shows ZnO nanostructures prepared by chemical vapor deposition utilizing zinc acetate as the evaporation source. It is quite interesting to note that what we observe under the microscope has many similarities to things around us in our daily life. Figure 10.3e is a photo of a sand beach taken at Toffino, BC, Canada, which shows great similarity to the ion beam textured Si substrates. Figure 10.3f shows a photo of sea urchins taken at Jagalchi market, Busan, Korea. Again, the sea urchins are very similar to ZnO nanostructures shown in Figure 10.3d, but much larger in size.

Another approach to prepare ZnO nanowires is by thermal oxidation of metallic zinc. Preparation of metal oxide nanowire by thermal oxidation was first reported by Tagagi [13] where microwhisker α-Fe_2O_3 was found by thermal oxidation of iron at 400–700°C. Tagagi [13] noticed that the length of the nanowire and oxidation time has the following relationship:

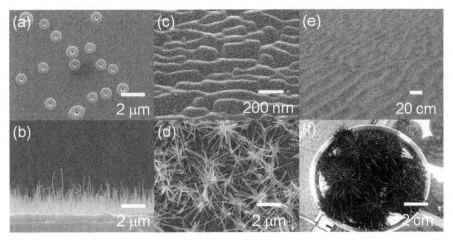

Figure 10.3 (a) ZnO nanodonuts, (b) ZnO nanowires, (c) ZnO QDs on ion beam textured Si substrates, (d) ZnO nanoclusters, (e) sand beach at Toffino, BC, Canada, and (f) photo of sea urchins, Jagalchi market, Busan, Korea.

$$L = kT^{0.3}, \qquad\qquad (10.1)$$

where L is the length of the nanowire, T is the oxidation time, and k is a temperature-dependent constant. Jiang et al. [14] reported that by simply heating TEM copper meshes in air results in the formation of CuO nanowire. Both the diameter and length of the CuO nanowire can be controlled by changing oxidation time and oxidation temperature, while the growth mechanism of CuO was later being identified due to diffusion [15] of copper across the Cu–Cu$_2$O interface. ZnO was found to exhibit a similar effect as that of CuO [16–22]. Kim et al. [16] demonstrated that thermal oxidation of metallic Zn nanoplates at 500°C results in the growth of single crystalline ZnO nanowires along the $\langle 11\bar{2}0 \rangle$ direction. In Kim's experiment, metallic zinc nanoplates were deposited on CaF$_2$ substrates by thermal evaporation and no catalysts were used in this process. PL study shows that the ZnO nanowire shows excellent near band edge (NBE) emission with almost negligible defect related visible emission. Fan et al. [17] obtained ZnO nanowires also along the $\langle 11\bar{2}0 \rangle$ direction by thermal oxidation of metallic zinc microcrystals at 600°C. Since the oxidation temperature is higher than the melting point of zinc which is 419°C, molten zinc is likely to be found on the microcrystal which may act as a eutectic solvent for ZnO. Fan thus proposed that the growth mechanism is due to a combination of liquid-solid and vapor-solid process. However, it is quite interesting to note that PL study shows that besides the NBE emission near 380 nm, strong defect-related emission near 540–610 nm was also observed. The defect-related green emission is likely due to the presence of oxygen vacancies [23]. Ren et al. [18] reported the growth of ZnO nanowires by thermal oxidation of metallic zinc at 400°C. A metallic zinc block was used in Ren's experiment and the surface of the metallic zinc block was polished by sand paper to remove surface oxidants and subsequently washed in dilute hydrochloric acid and de-ionized water before thermal oxidation. The ZnO nanowires obtained by Ren were also found to be single crystalline growing along the $\langle 11\bar{2}0 \rangle$ direction. However, PL study shows there is no NBE emission at all. Instead, only a strong defect-related emission centered at 400 nm was observed. This blue emission is likely due to interstitial zinc-related defects [24]. Since ZnO has a larger molar volume than zinc, during oxidation, a protective oxide layer will form on top of metallic zinc [25] and prevent the metallic zinc underneath to be further oxidized. Besides, the oxidation temperature used in Ren's experiment is below the melting point of zinc. Ren thus proposed that diffusion of atomic oxygen or zinc should be responsible for the formation of ZnO nanowires. Liu [19] prepared

Zn films by a reverse pulse plating process utilizing ZnCl and KCl as the main ingredient in their electrolyte. Oxidation of the platted zinc was performed at 300–600°C. After oxidation, ZnO nanowires were found across the zinc film and their field emission properties were compared. Law et al. [20] deposited patterned zinc film by lift-off process and thermal evaporation. After thermal oxidation, PL study shows strong NBE emission and negligible defect-related emission. The mechanism of ZnO nanowire prepared by thermal oxidation is further studied by Guo et al. [21] and Rackauskas et al. [22]. Guo prepared hexagonal-shaped metallic zinc disks by thermal evaporation. Guo's result shows that the {0001} surfaces of the zinc disk is stable and does not result in growth of ZnO nanowires after thermal oxidation. Instead, sublimation of zinc from the {0001} surfaces results in deposition of ZnO nanoneedles along the side of the hexagonal shaped zinc disks. PL study [21] shows that ZnO nanoneedles prepared by thermal oxidation at 400°C exhibit the strongest NBE emission with almost negligible defect-related emission. Rackauskas prepared zinc film by heating zinc wires under ambient air conditions from 400–900°C. They concluded that ZnO nanowire growth is determined by diffusion of zinc vacancies or interstitials.

10.4 Synthesis of ZnO Nanostructures Utilizing Ion Beam Processing

Both top–down and bottom–up approaches were used in the synthesizing of ZnO nanostructures utilizing ion beam processing. In the top–down approach, zinc nanocones were fabricated on zinc foils by ion beam sputtering. The aspect ratio of the zinc nanocone was found to be dependent on the temperature of the zinc foil. The formation mechanism of the nanocone is due to the presence of impurities or grain boundary defects that act as etch stop. A maximum aspect ratio of 25–30 can be achieved by ion beam in-situ heating and ion beam sputtering of the zinc foil. Thermal oxidation results in out-growth of ZnO nanowires from the shank of the nanocones.

In the bottom–up approach, ZnO QDs were deposited on ion beam textured Si substrates. Si substrates were first sputtered with argon ion beam with various beam incident angles. Quasi periodic nanoscale patterns were created on the Si substrate due to curvature dependent sputter rate. ZnO QDs were then deposited by reactive ion beam sputter deposition on the ion beam textured Si substrates. Beside ZnO QDs, ZnO nanowires were fabricated by thermal oxidation of metallic zinc films. The metallic zinc films were prepared by ion beam sputter deposition, RF magnetron sputtering and thermal evaporation.

Our results show that ZnO nanowires prepared by ion beam processing and thermal oxidation at 450°C exhibit excellent PL and photosensing properties. At low temperature (10 K), PL emission of ZnO nanowire is dominated by the recombination of surface exciton, while at room temperature, its PL is dominated by the recombination of free exciton. ZnO nanowire photosensors are fabricated by thermal oxidation and FIB micromachining which shows high photoconductive gain and fast photo response. Comparing the PL properties with that reported by other groups, the excellent PL and photosensing properties is attributed to the reduced impurities in the zinc film and low thermal oxidation temperatures.

10.5 Experiment and Results

Except FIB micro-machining, all the ion beam processing was performed utilizing the capillaritron ion source developed by our group. FIB micro-machining was carried out on a FEI Quanta 3D FEG Dual Beam system. For comparison, both thermal evaporation and RF magnetron sputtering were used for the deposition of metallic zinc films. The surface morphology of ZnO nanowires was investigated by a field emission scanning electron microscope (FE-SEM, JEOL JSM-6500F) operating at 15 kV. PL study was performed utilizing a 35-mW He–Cd laser at 325 nm as the excitation light source and the spectra were dispersed by a Horiba Jobin Yvon iHR 550 imaging spectrometer and detected by a Symphony CCD detector cooled to −70°C.

10.5.1 Ion Beam Synthesis of Zn Nanoneedles

Metallic zinc nanocones were fabricated by Ar^+ ion beam bombardment of zinc foils with beam energy from 4 to 10 keV (Figure 10.4). It is clear that as beam energy increases, the metallic zinc nanocone aspect ratio increases as well. The formation of the nanocone is due to impurities or grain boundary defect that acts as etch stop. As ion beam impinging on the zinc foil, curvature dependent sputtering rate results in the formation of cone-shaped nanostructure [26]. The aspect ratio of the nanocone increases as beam energy increases, which is due to increased ion beam in-situ heating [27]. As the beam energy increases from 4 to 10 keV, temperature of the zinc foil increases from 70 to 125°C [28]. Raising temperature of the zinc foil results in increased surface diffusion and thus nanocone with higher aspect ratio is achieved. This is confirmed by positioning the zinc foil on a heat sink during ion beam

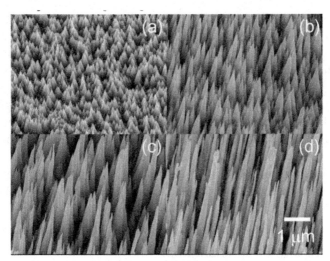

Figure 10.4 ZnO nanocone fabricated by ion beam sputtering utilizing a capillaritron ion source with beam energy of (a) 4, (b) 6, (c) 8, and (d) 10 keV.

sputtering. Under this condition, the aspect ratio of the nanocone only increases slightly from 1 to 2.3 as anode voltage increases from 4 to 10 keV [27, 28].

10.5.2 ZnO QDs on Ion Beam Textured Si Substrates

Ion beam textured substrates were prepared by capillaritron argon ion beam sputtering at various beam incident angles. Ion beam sputtering results in the formation of nanoscale ripples with a quasi wavelength from 70 to 150 nm [12]. ZnO QDs were then deposited by reactive ion beam sputter deposition utilizing the capillaritron ion source. During the deposition, both argon and oxygen were passed simultaneously through the ion source to act as sputtering and reacting gases, respectively. ZnO deposited at 250°C from 15 to 30 min results in the formation of QDs with diameter from 17.0 to 31.1 nm and height from 2.1 to 4.2 nm [12]. PL study shows that as QD height drops, PL peak position shifts from 3.33–3.41 eV. Since the QD exhibits an ellipsoidal shape, excitons may experience strong confinement in the vertical direction while weak or no confinement along the horizontal direction. The trend of ZnO QD PL peak position change is similar to that of quantum wells, indicating that these ZnO QDs act as quantum wells with a fractional dimension $\alpha_f = 2$. Further study shows that the ion beam textured Si substrates provides a wider processing window, better QD height control, and better QD density control [29].

As ion beam incident angle changes, quasi periodic ripple wavelength changes as well. The ripple wavelength was found to be dependent on beam energy (E) as $\propto E^{1.2-1.3}$, indicating that ion beam induced diffusion is the dominate diffusion mechanism for the formation of nanoscale ripples. Figure 10.5 shows the RMS roughness of ion beam textured Si substrate as a function of ion beam incident angles. As ion beam incident angle increases, RMS roughness increases to 11.2 nm as incident angle reaches 40° and then decreases. The change in RMS is due to the competition of quasi periodic ripples with wave vectors parallel and perpendicular to the projection of the incident ion beam. The formation mechanism of the nanoscale ripple is due to the balance between curvature-dependent sputtering rate and thermally induced diffusion [29].

10.5.3 ZnO Nanowire by Thermal Oxidation of Metallic Zinc

10.5.3.1 Zn foil with ion implantation

ZnO nanowires were prepared by thermal oxidation of metallic zinc foils or films. First, a metallic zinc foil was cleaned by ion beam sputtering to remove surface contamination. This top contaminated layer on a zinc foil acts as a diffusion barrier and should be removed to expose a clean surface. After ion beam cleaning, the zinc foil is positioned in a tube furnace heated at 450°C in flowing oxygen ambient for three hours. For comparison, ion implantation was performed on the zinc foil right after ion beam cleaning

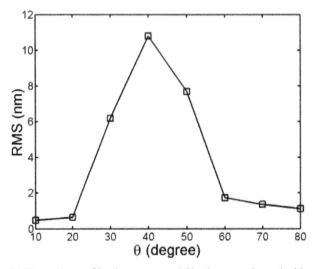

Figure 10.5　RMS roughness of ion beam textured Si substrate at beam incident angles from 10 to 80°. The argon ion beam energy is 8 keV.

but before thermal oxidation. Both oxygen and nitrogen implantation were performed to study the effect of impurities in zinc on the properties of ZnO nanowires.

Figure 10.6a shows a low-resolution SEM micrographs of a zinc foil after ion beam cleaning. The contrast is due to grains with different orientation that exhibit different secondary electron yields. Figure 10.6b shows a low-resolution SEM micrograph of such a zinc foil after thermal oxidation. The higher contrast is due to the formation of ZnO on the zinc foil. Figure 10.6c shows a high-resolution SEM micrograph of a zinc foil after thermal oxidation. showing the presence of ZnO nanowires across the film. Figure 10.6d shows a high-resolution SEM micrograph of a zinc foil after nitrogen implantation and thermal oxidation. Comparing with Figure 10.6c, it is interesting to note that nitrogen implantation results in enhanced ZnO nanowire density. TEM analysis shows that all the ZnO nanowires were grown along the $\langle 11\bar{2}0 \rangle$ with single crystalline quality [30]. However, their PL properties are quite different. At 10 K, both unimplanted and nitrogen-implanted samples shows NBE emission at 3.363 eV with similar full-width at half-maximum (FWHM) of $\sim$10 meV and negligible defect-related emission. Oxygen implanted sample, however, shows a NBE emission at 3.342 eV with a FWHM of 60 meV

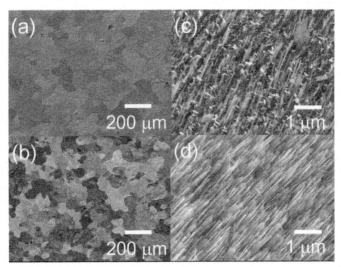

Figure 10.6 SEM micrographs of metallic zinc foils, (a) after ion beam cleaning and before thermal oxidation, (b) after ion beam cleaning and thermal oxidation, (c) same as (b) but at higher resolution showing the presence of ZnO nanowires, and (d) after ion beam cleaning, nitrogen ion implantation, and thermal oxidation.

and strong defect-related emission at 1.75–2.75 eV. Variable temperature PL study shows that the NBE emission of nitrogen implanted and unimplanted sample is due to the recombination of excitons or excitons bound to neutral donors, while that the NBE of oxygen implanted sample is due to defect-related emission. This indicates that the impurity in the zinc foil has a strong impact on ZnO nanowire PL properties. In this case, the impurity is oxygen, which reacts with zinc during thermal oxidation that results in the change of PL property [30]. TEM EDS analysis of oxygen-implanted sample shows a higher concentration of zinc. Both PL study and TEM EDS analysis indicates that ZnO nanowire obtained from oxygen-implanted zinc foils contains excess zinc. Excess zinc acts as donor in ZnO that results in higher carrier concentration and reduced work function. It is not surprised to see that oxygen-implanted zinc foils result in the formation of ZnO nanowire with the lowest threshold electric field in field emission characterization. Comparing these PL results with those previously reported by other research groups, it clearly suggests that contaminants in the zinc substrate have a significant impact on the properties of ZnO nanowires prepared by thermal oxidation.

10.5.3.2 Zn films deposited by ion beam sputter deposition

Metallic zinc films were also prepared by ion beam sputter deposition utilizing a capillaritron in source. A metallic zinc target (99.99%) was placed at 30 mm downstream of the capillaritron ion source, while Si substrates were placed at close proximity to the zinc target. The vacuum chamber was pumped down to 3×10^{-6} Torr by a turbo molecular pump. The deposition was performed with beam energy of 8, 10, and 16 keV. After the deposition, the zinc film is thermally annealed at 400°C for 3 hours in flowing oxygen ambient [31]. After thermal oxidation, ZnO nanowires were found across all the samples, except for that deposited with beam energy of 8 keV. This is due to the fact that the deposition rate is low and during the deposition of zinc films, residual oxygen in the vacuum chamber is mixed with the zinc film that result in no growth of ZnO nanowires after thermal oxidation. Figure 10.7 shows room temperature PL spectra of all the sample after thermal oxidation. From Figure 10.7, PL spectrum obtained from samples deposited with beam energy of 8 keV shows strong defect-related emission centered at 510 nm and weak near band edge emission centered at 373 nm. However, PL from samples deposited with beam energies at 12 and 16 keV show strong NBE centered at 375 nm and weak defect-related deep level emission at 500 nm. Also from Figure 10.7, it is clear that the emission intensity from 8 keV prepared sample is much weaker,

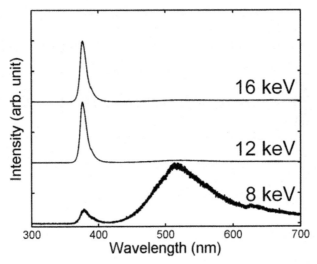

Figure 10.7 PL spectra of thermally oxidized samples deposited with beam energy at 8, 12 and 16 keV.

which can be seen from the low much lower signal-to-noise ratio, indicating that the NBE of ZnO nanowires is much stronger than that of the underlying ZnO films.

10.5.3.3 Zn films deposited by RF magnetron sputtering

Metallic zinc films were deposited on Si substrates at room temperature by RF magnetron sputtering utilizing a metallic zinc target (99.99%). Si substrates were positioned at 65 mm upstream of the zinc target. The base pressure and working pressure were 5×10^{-5} and 5×10^{-3} Torr, respectively. The discharge power were 70, 120, and 180 W. A mechanical shutter was positioned between the substrate and the target. The zinc target was presputtered, while the plasma emission spectra were monitored. Figure 10.8 shows plasma spectra taken during the presputtering and during the deposition of zinc films. The mechanical shutter remain closed until the plasma emission spectrum is dominated by zinc atomic ion emission [32].

Figure 10.9 shows SEM micrographs of zinc films deposited with discharge power of 70, 120, and 80 W. From Figure 10.9a, it shows that zinc film deposited by 70 W is porous, composed of partially oxidized fibers, while zinc film deposited by 120 W results in a dense film. Even though XRD shows no diffraction peaks of ZnO before thermal oxidation, the 70 W deposited zinc film results in no formation of ZnO nanowire after thermal oxidation. This is

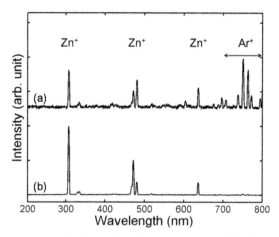

Figure 10.8   Plasma spectra of taken during (a) presputtering and (b) deposition of zinc films.

Figure 10.9   SEM micrographs of zinc films deposited by RF magnetron sputtering with discharge power of (a) 70 W, (b) 120 W, and (c) 180 W.

also attributed to the low deposition rate that residual oxygen in the deposition chamber is mixed with the zinc film. Thermal oxidation of 120 W deposited samples results in vertically aligned ZnO [32]. The length of ZnO nanowires obtained by oxidation of 120 W deposited samples is dependent on oxidation time as

$$L = 0.27T^{1/2}, \tag{10.2}$$

where L is the length of the ZnO nanowire in units of microns and T is the oxidation time in unit of minutes. The length dependence on oxidation time concludes that the formation mechanism of ZnO nanowire is due to diffusion. Thermal oxidation of 180 W deposited sample also results in the formation of ZnO nanowire. However, the nanowires are oriented in an angle relative to the substrate. This is attributed to the fact [21] that {0001} facets of zinc is more oxidation-resistant that ZnO nanowire is formed due to sublimation of zinc from the {0001} facets via vapor solid mechanism.

10.5.3.4 Zn films deposited by thermal evaporation

From the previous study, it shows that residual oxygen in the zinc films may hinder the growth of ZnO nanowire and result in the formation of interstitial zinc defects. To increase the deposition rate, thermal evaporation is used to deposit zinc films. In this study, zinc shots 3 mm in diameter were placed in a quartz crucible. SiO_2/Si substrates were positioned at 65 mm upstream of the zinc shot. The quartz crucible was heated by resistive heating to 500–600°C. A mechanical shutter was positioned between the substrate and the zinc evaporation source to control deposition time. The deposition was performed at a base pressure of 5×10^{-5} Torr with a deposition rate of 0.5 μm/min. For PL study, ZnO nanowires were prepared by thermal oxidation of zinc films. For photosensing study, a shadow mask with an opening of 350 μm was used to deposit metallic zinc stripes. A trench with a width of 3 μm was cut across the metallic Zn strip using FIB. Subsequent thermal oxidation results in growth of ZnO across the two ends of the Zn strips [33]. SEM micrographs shows ZnO nanowires were grown across the trench (Figure 10.10). The density of the nanowire is roughly 500 wires/mm. As a voltage of 0.1 V is applied across the nanowire detector, excellent photosensing properties were achieved with fast response time.

Variable temperature PL study shows that the room temperature PL is dominated by the recombination of free excitons, while 10 K PL is dominated by surface exciton, confirmed by power dependent PL study [33]. ZnO

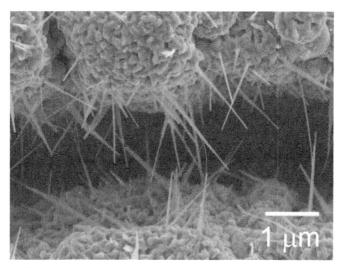

Figure 10.10   ZnO nanowires grown across the trench after thermal oxidation.

nanowire sensors were thus fabricated and exceptional sensing properties have been achieved.

10.6 Conclusions

A capillatron ion source has been developed and is used for surface cleaning and ion implantation. Metallic zinc nanocone is formed by ion beam sputtering. The aspect ratio of the nanocone is controlled by controlling the temperature of the zinc foil. Ion beam textured substrates provide a better processing window for the deposition of ZnO QDs. At least two mechanisms are involved in the growth of ZnO nanowires by thermal oxidation of metallic zinc. Diffusion of zinc across zinc-ZnO interface through grain boundary defects results in ZnO nanowires. The diffusion of zinc is highly likely to be grain orientation dependent. Sublimation of zinc from oxidation-resistant surfaces also results in the formation of ZnO nanowire via vapor-solid mechanism. Oxidation temperature has a great impact on the properties of ZnO nanowires. Oxidation temperatures close to the melting point of zinc result in high-quality ZnO nanowires with strong NBE emission and negligible defect-related emission. Higher oxidation temperatures result in strong oxygen vacancy defect-related green emission which is due to sublimation of ZnO. The ZnO nanowire property is also dependent on impurities in the zinc films. Should the zinc film contain oxygen, it may result in the formation of zinc interstitial defects or no ZnO nanowire growth at all. By reducing the impurities in zinc films, ZnO nanowire photosensors show high photoconductive gain, fast photoresponse, and excellent PL properties.

References

[1] J. F. Mahoney, J. Perel, and A. T. Forrester, Appl. Phys. Lett., 38 (1981) 320.

[2] J. Orloff, M. Utlaut, and L. Swanson, "High resolution focused ion beams: FIB and its applications" Springer Science+Business Media, LLC, New York, 2003.

[3] L. C. Chao and J. Orloff, J. Vac. Sci. Technol. B 15 (1997) 2732.

[4] L. C. Chao, B. K. Lee, C. J. Chi, J. Cheng, I. Chyr, and A. J. Steckl, Appl. Phys. Lett., 75 (1999) 1833.

[5] L. C. Chao and A. J. Steckl, Appl. Phys. Lett., 74 (1999) 2364.

[6] V. N. Tondare, J. Vac. Sci. Technol. A 23 (2005) 1498.

[7] Y. W. Heo, D. P. Norton, L. C. Tien, Y. Kwon, B. S. Kang, F. Ren, S. J. Pearton, and J. R. LaRoche, Materials Science and Engineering R 47 (2004) 1.

[8] B. D. Yao, Y. F. Chan, and N. Wang, Appl. Phys. Lett. 81 (2002) 757.

[9] C. Soci, A. Zhang, B. Xiang, S. A. Dayeh, D. P. R. Aplin, J. Park et al., Nano Lett. 7 (2007) 1003.

[10] L. C. Chao, P. C. Chiang, S. H. Yang, J. W. Huang, C. C. Liau, J. S. Chen et al., Appl. Phys. Lett. 88 (2006) 25111.

[11] L. C. Chao, S. H. Yang, Appl. Surf. Sci. 253 (2007) 7162.

[12] L. C. Chao, Y. K. Li, and W. C. Chang, Mater. Lett. 65 (2011) 1615.

[13] R. Takagi, J. Phys. Soc. Jpn. 12 (1957) 1212.

[14] X. Jiang, T. Herricks, and Y. Xia, Nano Lett. 2 (2002) 1333.

[15] A. M. Gonçalves, L. C. Campos, A. S. Ferlauto, and R. G. Lacerda, J. Appl. Phys. 106 (2009) 034303.

[16] T. W. Kim, T. Kawazoe, S. Yamazaki, M. Ohtsu, and T. Sekiguchi, Appl. Phys. Lett. 26 (2004) 3358.

[17] H. J. Fan, R. Scholz, F. M. Kolb, and M. Zacharias, Appl. Phys. Lett. 85 (2004) 4142.

[18] S. Ren, Y. F. Bai, J. Chen, S. Z. Deng, N. S. Xu, Q. B. Wu et al., Mater. Lett. 61 (2007) 666.

[19] Y. Liu, C. Pan, Y. Dai, and W. Chen, Mater. Lett. 62 (2008) 2783.

[20] J. B. K. Law, C. B. Boothroyd, and J. T. L. Thong, J. Crys. Growth 310 (2008) 2485.

[21] C. F. Guo, Y. Wang, P. Jiang, S. Cao, J. Miao, Z. Zhang, and Q. Liu, Nanotechnology 19 (2008) 445701.

[22] S. Rackauskas, A. G. Nasibulin, H. Jiang, Y. Tiang, G. Statkute, S. D. Shandakov et al., Appl. Phys. Lett. 95 (2009) 183114.

[23] M. Li, G. Xing, G. Xing, B. Wu, T. Wu, X. Zhang et al., Phys. Rev. 87 (2013) 115309.

[24] H. Zeng, G. Duan, Y. Li, S. Yang, X. Xu, and W. Cai, Adv. Funct Mater. 20 (2010) 561.

[25] N. Cabrera and N. F. Mott, Rep. Prog. Phys. 12 (1949) 163.

[26] C. H. Hsu, H. C. Lo, C. F. Chen, C. T. Wu, J. S. Hwang, D. Das et al., Nano Lett. 4 (2004) 471.

[27] L. C. Chao, C. C. Liau, S. J. Lin, and J. W. Lee, J. Vac. Sci. Technol. B 26 (2008) 2601.

[28] L. C. Chao, C. C. Liau, and J. W. Lee, US Patent 8,216,480 B2, 2012.

[29] L. C. Chao, W. R. Chen, J. W. Chen, S. M. Lai, and G. Keiser, J. Vac. Sci. Technol. B 29 (2011) 051805.

[30] L. C. Chao, J. W. Lee and C. C. Liu, J. Phys. D: Appl. Phys. 41 (2008) 115405.

[31] L. C. Chao, C. F. Lin, and C. C. Liau, Vacuum 86 (2011) 295.

[32] L. C. Chao, S. Y. Tsai, C. N. Lin, C. C. Liau, and C. C. Ye, Materials Science in Semiconductor Processing 16 (2013) 1316.

[33] L. C. Chao, C. C. Ye, Y. P. Chen, and H. Z. Yu, Appl. Surf. Sci. 282 (2013) 384.

Index

About the Editors

Shien-Kuei Liaw received doctorate degrees in electro-optical engineering and mechanical engineering from National Chiao-Tung University (Taiwan) and National Taiwan University (Taiwan) in 1999 and 2014, respectively. In 1993, he joined the Chung-Hua Telecommunication Institute, Taiwan. Since then, he has been working on optical communication and fiber-based technologies. He was a visiting researcher at Bellcore (now Telcordia), USA, in 1996 for six months and a visiting professor at Oxford University, UK, in 2011 for three months. He joined the Department of Electronic Engineering, National Taiwan University of Science and Technology (NTUST) in 2000. He has been director of Optoelectronics Research Center and Technology Transfer Center in NTUST. He has been an active member in the technical programs committee and has invited speakers and held session chairs for many international conferences. He has bagged quite a few national honors and awards, such as Outstanding Professor of the Chinese Institute of Electrical Engineering in 2015; the best project award of National Science and Technology Program for Telecommunication in 2006; the outstanding Youth Award of The Chinese Institute of Electrical Engineering; and the outstanding Youth Academic Award of the Optical Engineering Society of the Republic of China. Prof. Liaw has authored and coauthored over 250 international journal articles and conference presentations. He also serves as an associate editor for Fiber and Integrated Optics. Currently, Prof. Liaw is a distinguished professor of NTUST and the Taipei Chapter Chair, IEEE Photonics Society.

Gong-Ru Lin received B.S. degree in Physics from Soochow University in 1988 and M.S. and Ph.D degrees in electro-optical engineering from National Chiao Tung University (NCTU) in 1990 and 1996, respectively. He has ever been the faculty with National United University, Tatung University, National Taipei University of Technology, and National Chiao Tung University from 1997 to 2006. Since 2006, he joined the Graduate Institute of Photonics and Optoelectronics (GIPO), and also the jointly appointed professor with the Department of Electrical Engineering, National Taiwan University. Prof. Lin has a broad research spectrum covering ultrafast fiber lasers, semiconductor laser diodes, fiber-optic communications, and silicon and carbon nanophotonics. He was the Taipei Chapter Chair of IEEE Photonics Society. He served as the associate editor of "Journal of Lightwave Technology", "IEEE Photonics Journal" and "Current Nanoscience". He received outstanding research awards from the National Science Council in 1997, 1998, and 2000. In 2011, he received the three-year Distinguished Research Award from the National Science Council. To date, Prof. Lin has been promoted as a senior member of IEEE in 2004, the Fellow of SPIE (FSPIE) in 2008, the Fellow of IET (FIET) in 2009, the Fellow of IOP (FInstP) in 2010, and the Fellow of OSA in 2013. Currently, he is the OSA traveling lecturer and the SPIE visit lecturer. In 2015, he was elected as the distinguished professor of National Taiwan University.